GCSE

Chemistry

Complete Revision
and Practice

Contents

Contents

The Periodic Table

Periods

You should be pretty familiar with the 35 elements shown shaded.

You don't really need to know anything about the others.

Mass number →
Atomic number →

The Lanthanides

The Actinides

Common Ions You Really Should Know

1^+ ions	2^+ ions	3^+ ions	$4^+/4^-$	3^-	2^- ions	1^- ions
Li^+ (lithium)	Mg^{2+} (magnesium)	Al^{3+} (aluminium)	Very rare	Fairly rare	O^{2-} (oxide)	F^- (fluoride)
Na^+ (sodium)	Ca^{2+} (calcium)	Fe^{3+} (iron(III))			S^{2-} (sulphide)	Cl^- (chloride)
K^+ (potassium)	Ba^{2+} (barium)	Cr^{3+} (chromium(III))				Br^- (bromide)
Cu^+ (copper(I))	Cu^{2+} (copper(II))					I^- (iodide)
Ag^+ (silver)	Fe^{2+} (iron(II))	Note that copper and iron can both form two different ions.				NO_3^- (nitrate)
H^+ (hydrogen)	Zn^{2+} (zinc)				SO_4^{2-} (sulphate)	OH^- (hydroxide)
NH_4^+ (ammonium)	Pb^{2+} (lead)				CO_3^{2-} (carbonate)	HCO_3^- (hydrogencarbonate)
These atoms lose *one* electron to form 1⁺ ions.(NH_4^+ isn't an atom)	These atoms lose *two* electrons to form 2⁺ ions	These atoms lose *three* electrons to form 3⁺ ions	Atoms find it very hard to gain or lose three or four electrons.		These atoms/molecules gain *two* electrons to form 2⁻ ions	These atoms/molecules gain *one* electron to form 1⁻ ions

Published by Coordination Group Publications Ltd

Editors:
Richard Parsons, Alice Shepperson.

Contributors:
Ruth Amos, Martin Chester, Dominic Hall, Gemma Hallam, Phillipa Hulme, Munir Kawar, Diana Lazarewicz, Becky May, Mike Thompson, James Paul Wallis, Jim Wilson, Suzanne Worthington.
Illustrations by Sandy Gardner   e-mail: illustrations@sandygardner.co.uk.   With thanks to Jeremy Cooper for the proofreading.

AQA (NEAB)/AQA examination questions are reproduced by permission of the Assessment and Qualifications Alliance.
Edexcel (London Qualifications Ltd) examination questions are reproduced by permission of London Qualifications Ltd.
OCR examination questions are reproduced by permission of OCR.

Please note that AQA questions from pre-2003 papers are from legacy syllabuses and not from the current specification.
CGP has carefully selected the questions contained in the practice exam to cover subject areas which are still relevant to the current specification. As such the exam provides a good test of your knowledge of the syllabus areas you need to understand to get a good grade in the real thing.

ISBN-10: 1 84146 371 X
ISBN-13: 978 1 84146 371 1
Website: www.cgpbooks.co.uk
Printed by Elanders Hindson Ltd, Newcastle upon Tyne.
Clipart source: CorelDRAW®

Solids, Liquids and Gases

These are known as the <u>three states of matter</u>. Make sure you know everything there is to know.

Solids have **Strong Forces** of Attraction

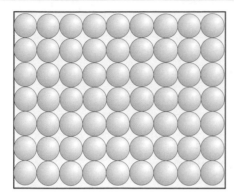

1) There are <u>strong forces</u> of attraction between molecules.
2) The molecules are held in <u>fixed positions</u> in a very regular <u>lattice arrangement</u>.
3) They <u>don't move</u> from their positions, so all solids keep a <u>definite shape</u> and <u>volume</u>, and don't flow like liquids.
4) They <u>vibrate</u> about their positions. The <u>hotter</u> the solid becomes, the <u>more</u> they vibrate. This causes solids to <u>expand</u> slightly when heated.
5) Solids <u>can't</u> be <u>compressed</u> because the molecules are already packed <u>very closely</u> together.
6) Solids are generally <u>very dense</u>.

Liquids have **Moderate Forces** of Attraction

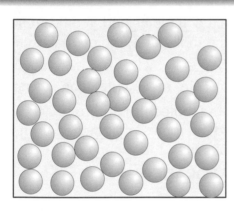

1) There is <u>some force</u> of attraction between the molecules.
2) The molecules are <u>free</u> to <u>move</u> past each other, but they do tend to <u>stick together</u>.
3) Liquids <u>don't</u> keep a <u>definite shape</u> and will flow to fill the bottom of a container. But they do keep the <u>same volume</u>.
4) The molecules are <u>constantly</u> moving in <u>random motion</u>. The <u>hotter</u> the liquid becomes, the <u>faster</u> they move. This causes liquids to <u>expand</u> slightly when heated.
5) Liquids <u>can't</u> be <u>compressed</u> because the molecules are already packed closely together.
6) Liquids are quite dense, but not as dense as solids.

Gases have **No** Forces of Attraction

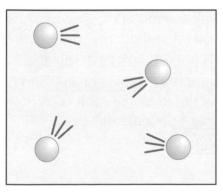

Pressure is exerted by molecules bouncing off the walls of the container.

1) There is <u>no force</u> of attraction between the molecules.
2) The molecules are <u>free</u> to <u>move</u>. They travel in <u>straight lines</u> and only interact with each other <u>when they collide</u>.
3) Gases <u>don't</u> keep a definite <u>shape</u> or <u>volume</u> and will always <u>expand to fill</u> any container. Gases exert a <u>pressure</u> on the walls of the container.
4) The molecules are <u>constantly</u> moving in <u>random motion</u>. The <u>hotter</u> the gas becomes, the <u>faster</u> they move. When <u>heated</u>, a gas will either <u>expand</u> or its <u>pressure</u> will <u>increase</u>.
5) Gases can be <u>compressed</u> easily because there's <u>a lot of free space</u> between the molecules.
6) Gases all have <u>very low densities</u>.

More energy makes the molecules move more

This is pretty basic stuff, but people still lose marks in the Exam because they don't learn all the little details. Make sure you know it. Try covering the page and scribbling down all the facts from memory.

2

Changes of State

CHANGES OF STATE always involve <u>HEAT ENERGY</u> going either <u>IN</u> or <u>OUT</u>.

Melting — the *rigid lattice* breaks down

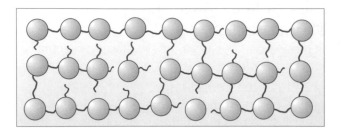

1) When a <u>SOLID</u> is <u>heated</u>, the heat energy goes to the molecules.

2) It makes them vibrate <u>more and more</u>.

3) Eventually <u>the strong forces</u> between the molecules (that hold them in the rigid lattice) are <u>overcome</u>, and the molecules start to <u>move around</u>.

This is <u>MELTING</u>.

Evaporation — the *fastest* molecules *escape*

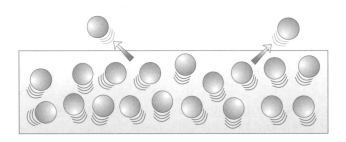

1) When a <u>LIQUID</u> is <u>heated</u>, the heat energy goes to the molecules, which makes them <u>move faster</u>.

2) Some molecules move faster than others.

3) Fast-moving molecules <u>at the surface</u> will <u>overcome</u> the <u>forces of attraction</u> from the other molecules and <u>escape</u>.

This is <u>EVAPORATION</u>.

Boiling — *all* molecules are *fast* enough to *escape*

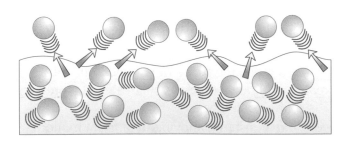

1) When the liquid gets <u>hot enough</u>, virtually <u>all</u> the molecules have enough speed and energy to overcome the forces and <u>escape each other</u>.

2) At this point big <u>bubbles of gas</u> form inside the liquid as the molecules <u>break away</u> from each other.

This is <u>BOILING</u>.

Evaporation is NOT the same as boiling

These changes of state are all caused by heating. The key idea here is that molecules move more when they're hot, and try to break away from each other. In the Exam, you'll need to talk about this increase in movement, and how it allows molecules to break the forces of attraction between them.

2

Changes of State

Changes of state can be seen as <u>flat spots</u> on heating and cooling <u>graphs</u>.

Heating and Cooling Graphs Have **Important Flat Spots**

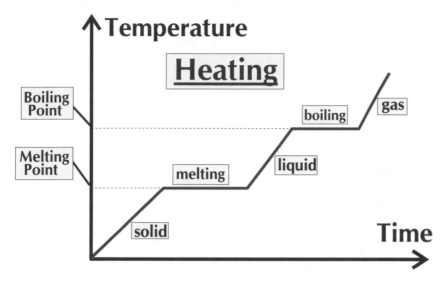

1) When a substance is <u>MELTING</u> or <u>BOILING</u>, all the <u>heat energy</u> supplied is used for <u>breaking bonds</u> rather than raising the temperature, hence the <u>flat spots</u> in the heating graph.

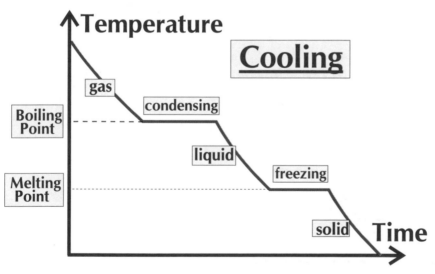

2) When a <u>liquid</u> is <u>cooled</u>, the graph for temperature will show a <u>flat spot</u> at the <u>freezing point</u>.
3) As the molecules <u>fuse</u> into a solid, <u>HEAT IS GIVEN OUT</u> as the <u>bonds form</u>, so the temperature <u>won't</u> go down until <u>all</u> the substance has turned to <u>solid</u>.

Learn what the flat spots mean

Energy is needed to break the bonds between molecules. The opposite is also true — making bonds produces energy. You get flat spots on cooling graphs because the heat produced by bonds forming during a change of state cancels out the energy being lost to the environment.

Warm-Up and Worked Exam Questions

Now you need to practise what you've learnt about changes of state.
Have a go at these easy warm-up questions to get you started.

Warm-up Questions

1) What are the three states of matter?
2) With reference to the spacing of molecules, explain why gases can be compressed and solids cannot be compressed.
3) Describe how water particles evaporate from the surface of a beaker of water.
4) Explain what happens to water particles when ice melts.

Worked Exam Questions

Now look over this worked example. It'll give you an idea of how you should write down your answers in the Exam. Then try the exam questions.

1   Describe the forces of attraction between the molecules in a:
   i)   Solid

   The forces of attraction between molecules in a solid are strong.
   (1 mark)

   ii) Liquid

   Some forces of attraction but they're not as strong as in solids.
   (1 mark)

   iii) Gas

   There are no forces of attraction between gas molecules.
   (1 mark)

2
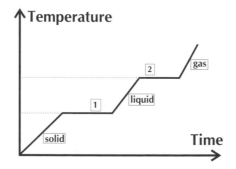

   Study the diagram above. Explain in terms of bonds the processes happening at:

   i)   **1** *Melting. The substance is changing from solid to liquid. Heat energy weakens intermolecular bonds, allowing molecules to move.*
   (2 marks)

   ii) **2** *Boiling. The substance is changing from liquid to gas. Heat energy breaks the intermolecular bonds completely.*

   This is why learning definitions is such a good idea.   *(2 marks)*

Exam Questions

1   By describing the movement of its atoms, explain what happens to a piece of solid iron when it is warmed.

....

....

(2 marks)

2   What effect does heating have on the movement of gas molecules?

....

(1 mark)

3

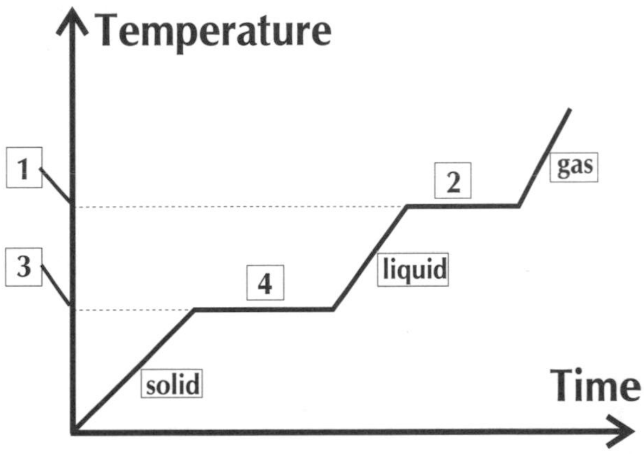

a)   On the above graph, which number represents the melting point of the substance?

....

(1 mark)

b)   Which number represents the time when the substance is melting?

....

(1 mark)

c)   During the transition from solid to liquid the temperature remains the same even though the substance is being heated. Explain this in terms of bonds between the molecules.

....

....

....

(3 marks)

Atoms

The structure of atoms is really simple. All you need to do is learn it.

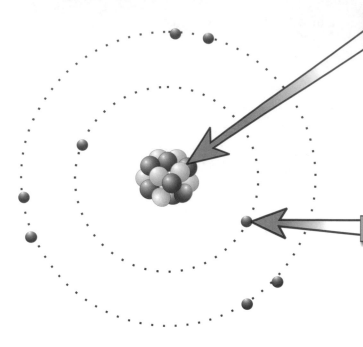

*Don't forget that atoms are <u>very small</u>.
They're <u>too small to see</u>, even with a microscope.*

The Nucleus

1) It's in the <u>middle</u> of the atom.
2) It contains <u>protons</u> and <u>neutrons</u>.
3) It has a <u>positive charge</u> because it contains protons, and protons are positively charged.
4) Almost the <u>whole mass</u> of the atom is <u>concentrated</u> in the <u>nucleus</u>.
5) But size-wise it's <u>tiny</u> compared to the atom as a whole.

The Electrons

1) Move <u>around</u> the nucleus.
2) They're <u>negatively charged</u>.
3) They're <u>tiny</u>, but they cover a <u>lot of space</u>.
4) The <u>volume</u> of their orbits determines <u>how big</u> the atom is.
5) They have <u>virtually no mass</u>.
6) They occupy <u>shells</u> around the nucleus.
7) These shells explain the <u>whole of Chemistry</u>.

Number of Protons **Equals** Number of Electrons

1) Neutral atoms have <u>no charge</u> overall.
2) The <u>charge</u> on the <u>electrons</u> is the <u>same size</u> as the charge on the <u>protons</u> but <u>opposite</u>.
3) This means the <u>number of protons</u> always <u>equals</u> the <u>number of electrons</u> in a <u>neutral atom</u>.
4) If some electrons are <u>added</u> or <u>removed</u>, the atom becomes <u>charged</u> and is then an <u>ion</u>.
5) The number of neutrons isn't fixed but is usually <u>just a bit higher</u> than the number of protons.

Know Your Particles

<u>Protons are Heavy</u> and <u>Positively Charged</u>
<u>Neutrons are Heavy</u> and <u>Neutral</u>
<u>Electrons are Tiny</u> and <u>Negatively Charged</u>

PARTICLE	MASS	CHARGE
Proton	1	+1
Neutron	1	0
Electron	$\frac{1}{2000}$	-1

Know your enemy...

The atom is the foot soldier of chemistry. You'll never conquer the trickier bits unless you understand the atom thoroughly. You MUST memorise this page or you'll be Exam cannon-fodder.

Atomic Number and Mass Number

These are just two simple numbers. All you've got to do is learn what they tell you about an atom.

The Mass Number —————— ^{23}

—The total number of Protons and Neutrons

The Atomic Number

— Number of Protons —————— $_{11}$ **Na**

Points to Note

1) The <u>atomic number</u> tells you how many <u>protons</u> there are.

2) This <u>also</u> tells you how many <u>electrons</u> there are.

3) To get the number of <u>neutrons</u> — just <u>subtract</u> the <u>atomic number</u> from the <u>mass number</u>.

4) The <u>mass number</u> is always the <u>biggest</u> number. It tells you the relative mass of the atom.

5) The <u>mass</u> number is always <u>roughly double</u> the <u>atomic</u> number.

6) Which means there's about the <u>same</u> number of protons as neutrons in any nucleus.

*Isotopes are the **same** except for an extra **neutron** or two*

A favourite Exam question: ***"Explain what is meant by the term Isotope"***

The trick is that it's impossible to explain what one isotope is.
You have to outsmart them and always start your answer *"ISOTOPES ARE..."*

<u>LEARN</u> this definition:

> **Isotopes are:** different atomic forms of the <u>same element</u>, which have the SAME number of PROTONS but a DIFFERENT number of NEUTRONS.

1) The upshot is: isotopes must have the <u>same atomic number</u> but <u>different mass numbers</u>.

2) <u>If</u> they had <u>different</u> atomic numbers, they'd be <u>different elements altogether</u>.

3) A very popular pair of isotopes are <u>carbon-12</u> and <u>carbon-14</u>.

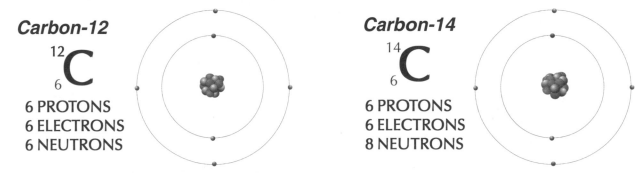

Carbon-12

$^{12}_{6}$**C**

6 PROTONS
6 ELECTRONS
6 NEUTRONS

Carbon-14

$^{14}_{6}$**C**

6 PROTONS
6 ELECTRONS
8 NEUTRONS

The number of electrons decides the chemistry of the element.
If the <u>atomic number</u> is the same, then the <u>number of protons</u> is the same, so the <u>number of electrons</u> is the same, so the <u>chemistry</u> is the same.
The different number of neutrons in the nucleus doesn't affect the chemical behaviour at all.

Chemistry by numbers

There really isn't that much information on this page — three definitions, a couple of diagrams and a dozen or so extra details. All you've got to do is <u>read it</u>, <u>learn it</u>, <u>cover the page</u> and <u>scribble it all down again</u>.

Atoms (Electron Shells)

The fact that electrons occupy "shells" around the nucleus is really important for the whole of chemistry. Remember that, and watch how electron shells are applied to each bit of it.

Electron Shell Rules:

1) Electrons always occupy <u>shells</u> or <u>energy levels</u>.

2) The <u>lowest</u> energy levels are <u>always filled first</u>.

3) Only <u>a certain number</u> of electrons are allowed in each shell:

<u>1st shell:</u>	2
<u>2nd Shell:</u>	8
<u>3rd Shell:</u>	8

4) Atoms are much <u>happier</u> when they have <u>full electron shells</u>.

5) In most atoms the <u>outer shell</u> is <u>not full</u> and this makes the atom want to <u>react</u>.

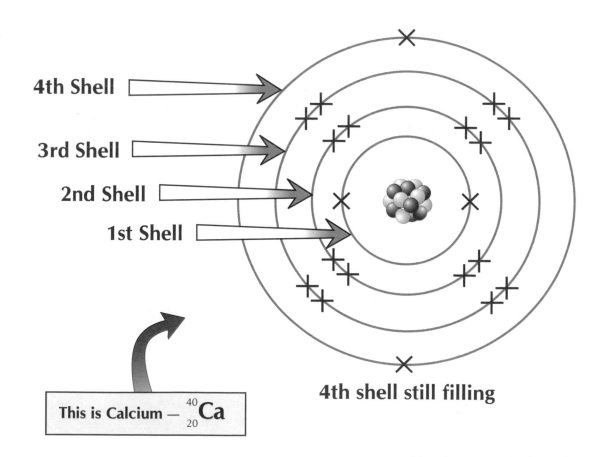

4th Shell

3rd Shell

2nd Shell

1st Shell

4th shell still filling

This is Calcium — $^{40}_{20}$Ca

It's all about filling shells

Remember, above all else, atoms want to have full electron shells. Full shells mean happy atoms that don't need to go and react with other atoms to gain or lose electrons. The only atoms with full electron shells are the noble gases, and you'd have a job getting them to react with anything.

Ionic Bonding

Ionic *Bonding* — *Swapping* Electrons

In <u>ionic bonding</u>, atoms <u>lose or gain electrons</u> to form <u>charged particles</u> (ions) which are then <u>strongly attracted</u> to one another (because of the attraction of opposite charges, + and –).

A shell with just **one** electron is **very keen to get rid of it**...

<u>All</u> the atoms over at the <u>left hand side</u> of the periodic table, such as <u>sodium, potassium, calcium</u> etc. have just <u>one or two electrons</u> in their outer shell.

Basically they're <u>keen to get rid of their outer electrons</u>, because then they'll only have <u>full shells</u> left, which is how they <u>like</u> it.

So when they get the chance, these atoms get rid of an electron or two, and become <u>ions</u> instead.

These ions have lost electrons, so they have a **+** charge.

This means they tend to <u>leap</u> at the first passing ion with an <u>opposite charge</u> and stick to it like glue.

A **nearly full** shell is keen to get that **extra electron**...

On the <u>other side</u> of the periodic table, the elements in <u>Group Six</u> and <u>Group Seven</u>, such as <u>oxygen</u> and <u>chlorine</u>, have outer shells which are <u>nearly full</u>.

They're obviously keen to <u>gain</u> the <u>extra one or two electrons</u> to fill the shell up.

When they do of course they become <u>ions</u> with a **-** charge, as they have more electrons than protons.

They're attracted to positive ions, and so latch on to the ion they've just taken electrons from, which is now positively charged. The reaction of sodium and chlorine is a <u>classic case</u>:

The <u>sodium</u> atom <u>gives up</u> its <u>outer electron</u> and becomes an Na⁺ ion.

The <u>chlorine</u> atom <u>picks up</u> the <u>spare electron</u> and becomes a Cl ⁻ ion.

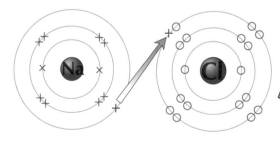

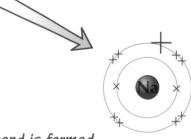

An <u>ionic bond</u> is formed.

Full Shells — it's the name of the game

This whole page of words and diagrams is just hammering home three very basic points:
1) Ionic bonds involve swapping electrons. 2) Some atoms like to lose electrons, some like to gain them.
3) Swapping electrons creates oppositely charged ions which are attracted to each other.

Covalent Bonding

Covalent Bonds — Sharing Electrons

1) <u>Sometimes</u> atoms prefer to make <u>covalent bonds</u> by <u>sharing electrons</u> with other atoms.
2) This way <u>both atoms</u> feel that they have <u>a full outer shell</u>, and that makes them happy.
3) <u>Each</u> covalent bond provides <u>one extra</u> shared electron for each atom.
4) Each atom involved has to make <u>enough</u> covalent bonds to <u>fill up</u> its outer shell.
5) <u>Learn</u> these <u>five</u> important examples, and the 'dot and cross' diagrams that go with them:

1) Chlorine Gas, Cl₂

Each chlorine atom needs just <u>one more</u> <u>electron</u> to complete the outer shell.

So they form <u>a single covalent bond</u> and together share <u>one pair</u> of electrons.

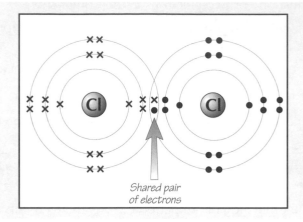

Shared pair of electrons

2) Hydrogen Gas, H₂

Hydrogen atoms have just one electron. They <u>only need one more</u> to complete the first shell.

So hydrogen atoms often form <u>single covalent bonds</u> to achieve this (as shown in four out of the six examples on this double page — see if you can spot which four).

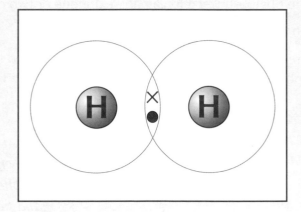

3) Ammonia, NH₃

Nitrogen has <u>five</u> outer electrons.

So it needs to form <u>three covalent bonds</u> to make up the extra <u>three</u> electrons needed.

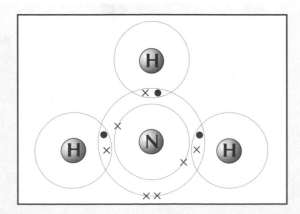

Covalent bonds are all about sharing

Make sure you know the first four points at the top of this page, and that you can use them to explain why atoms bond covalently. Remember, this is about sharing electrons here, not swapping them.

Covalent Bonding

4) Methane, CH$_4$

Carbon has <u>four outer electrons</u>, which is a <u>half full</u> shell.

To become a 4+ or a 4– ion is hard work, so it forms <u>four covalent bonds</u> to make up its outer shell.

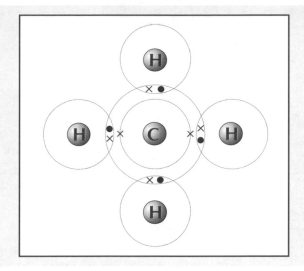

5) Oxygen Gas O$_2$, and Water H$_2$O

An <u>oxygen</u> atom has <u>six</u> outer electrons. Sometimes it forms <u>ionic</u> bonds by <u>taking</u> two electrons to complete the outer shell.

However, an oxygen atom will also cheerfully form <u>covalent bonds</u> and <u>share</u> two electrons instead, as in <u>oxygen gas</u>...

OXYGEN GAS

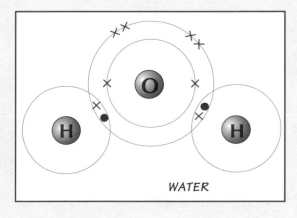

...or <u>water molecules</u>, where it <u>shares</u> electrons with the hydrogen atoms.

WATER

Learn the examples

Make sure you can draw all six molecules and explain exactly why they form the bonds that they do. <u>All from memory of course</u>.

Warm-Up and Worked Exam Questions

These warm-up questions should ease you in...

Warm-up Questions

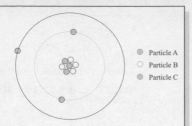

Particle A
Particle B
Particle C

1) The diagram on the right shows the arrangement of particles in an atom of the element lithium.
   Identify the particles A, B and C.

2) Using the examples of carbon-12 and carbon-14, explain what is meant by the term 'isotope'.

3) Using 'dot and cross' diagrams show the bonding in molecules of: chlorine gas (Cl_2), oxygen gas (O_2) and methane (CH_4).

Worked Exam Questions

If there were any warm-up questions you couldn't do, look back over the section and learn those bits again. Then have a look at the worked examples, and try the exam questions.

1   Write the electron arrangement for sodium Na (atomic number 11).

   2, 8, 1   *In a neutral atom the atomic number (11) is the same as the number of electrons. Electrons are arranged in energy levels according to the rule:* (1 mark)
   first shell: up to 2 — second shell: up to 8 —third shell: up to 8

2   Carbon-12 has atomic number 6 and mass number 12.
    Carbon-14 has mass number 14.

   a)   How many neutrons are there in the nucleus of a:

   i) Carbon-12 atom?

   12 - 6 = 6

   (1 mark)

   ii)Carbon-14 atom?

   14 - 6 = 8

   (1 mark)

   b)   Explain how carbon-12 and carbon-14 are isotopes of carbon.

   They both have the same number of protons (atomic

   number 6) but different numbers of neutrons (6 & 8)

   Know your definitions.

   (2 marks)

Exam Questions

1 a) What is an ion?

....

....

(2 marks)

  b) How many electrons does an Al^{3+} ion have? Aluminium has an atomic number of 13, and a mass number of 27.

....

(1 mark)

2 Sodium has the electron arrangement 2,8,1.
  Chlorine has the electron arrangement 2,8,7.

  a) Write the electron arrangement of:

    i) A sodium ion Na^+

    ....

    (1 mark)

    ii) A chloride ion Cl^-

    ....

    (1 mark)

  b) What type of bonding is found in sodium chloride?

  ....

  (1 mark)

  c) Using a 'dot and cross' diagram show the bonding between sodium and chlorine in sodium chloride.

  (2 marks)

  d) Using a 'dot and cross' diagram show the bonding between hydrogen and oxygen in water.

  (2 marks)

Ionic Substances

Simple Ions — Groups *1 & 2* and *6 & 7*

1) The elements that most readily form <u>ions</u> are those in Groups 1, 2, 6 and 7.

2) <u>Group 1 and 2 elements</u> are <u>metals</u> and they <u>lose</u> electrons to form <u>+ve ions</u> or <u>cations</u>.

3) <u>Group 6 and 7 elements</u> are <u>non-metals</u>. They <u>gain</u> electrons to form <u>–ve ions</u> or <u>anions</u>. Make sure you know these easy ones:

Cations		Anions	
Gr I	**Gr II**	**Gr VI**	**Gr VII**
Li⁺	Be²⁺	O²⁻	F⁻
Na⁺	Mg²⁺		Cl⁻
K⁺	Ca²⁺		

4) When any of the above elements <u>react together</u>, they form <u>ionic bonds</u>.

5) Only elements at <u>opposite sides</u> of the periodic table will form ionic bonds, e.g. Na and Cl, where one of them becomes a <u>cation</u> (+ve) and one becomes an <u>anion</u> (–ve).

> Remember, in sodium <u>metal</u> there are <u>only neutral sodium atoms</u>, Na.
> The Na⁺ ions <u>will only form</u> if the sodium metal <u>reacts</u> with something like water or chlorine.

Giant Ionic Structures don't melt easily, but when they do...

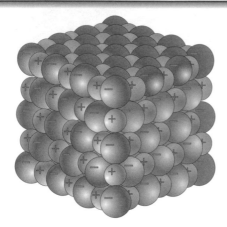

1) <u>Ionic bonds</u> always produce <u>giant ionic structures</u>.

2) The ions form a <u>closely packed regular lattice</u> arrangement.

3) There are <u>very strong ionic bonds</u> between <u>all</u> the oppositely charged ions.

4) A single crystal of salt is <u>one giant ionic lattice</u>, which is why salt crystals tend to be cuboid in shape:

1) They have **High melting points and boiling points**

Due to the <u>very strong</u> bonds between <u>all the ions</u> in the giant structure.

2) They **Dissolve** to form solutions that **conduct electricity**

When <u>dissolved</u> the ions <u>separate</u> and are all <u>free to move</u> in the solution, so obviously they'll <u>carry electric current</u>.

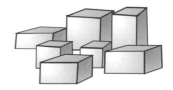

Dissolved in Water

Melted

3) They **Conduct** electricity when **molten**

When it <u>melts</u>, the ions are <u>free to move</u> and they'll carry electric current.

I hereby charge you with being an ion...

<u>Learn</u> which atoms form 1+, 1–, 2+ and 2– ions, and why (see page 9). Then learn all the features of ionic solids. When you think you know it all, <u>cover the page</u> and scribble down what you've learnt.

Covalent Substances: Two Kinds

Substances formed with <u>covalent bonds</u> can be either <u>simple molecules</u> or <u>giant molecular structures</u>.

Simple *Molecular* Substances

1) The atoms form <u>very strong covalent bonds</u>, creating <u>small molecules</u> of several atoms.

2) By contrast, the forces of attraction <u>between</u> these molecules are <u>very weak</u>.

3) The <u>result</u> of these <u>feeble intermolecular forces</u> is that the melting- and boiling-points are <u>very low</u>, because the molecules are <u>easily parted</u> from each other.

4) Most molecular substances are <u>gases or liquids</u> at room temperature.

5) Molecular substances <u>don't conduct electricity</u>, simply because there are <u>no ions</u>.

6) They <u>don't dissolve in water</u>, usually.

7) You can usually tell a molecular substance just from its <u>physical state</u>, which is always kind of "<u>mushy</u>" — i.e. <u>liquid</u> or <u>gas</u> or an <u>easily-melted solid</u>.

Very weak intermolecular forces

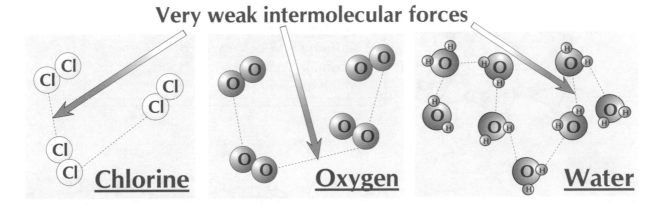

Chlorine   **Oxygen**   **Water**

Buckminster Fullerene

Fullerenes were discovered in 1985. They are really big carbon molecules, sometimes used as lubricants.

1) <u>60 carbon atoms</u> joined in a big ball.

2) Each carbon atom forms <u>three covalent bonds</u>.

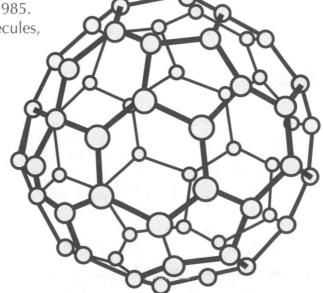

Know your simple molecular substances

Though you can tell a simple covalently bonded molecular substance by its "mushy" physical state, remember that ionic substances in solution, and warm metallic substances can also be somewhat runny.

Covalent Substances: Two Kinds

Giant Covalent *Structures*

1) These are similar to giant ionic structures except that there are <u>no charged ions</u>.
2) <u>All</u> the atoms are <u>bonded to each other</u> by <u>strong covalent bonds</u>.
3) They have <u>very high</u> melting and boiling points.
4) They <u>don't</u> conduct electricity — not even when <u>molten</u>.
5) They're usually <u>insoluble</u> in water.
6) The <u>main examples</u> are <u>diamond</u> and <u>graphite</u>, which are both made up of only <u>carbon atoms</u>.

Diamond

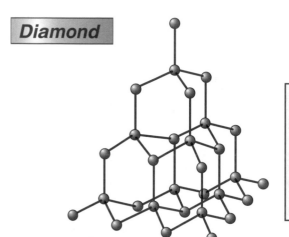

Each carbon atom forms <u>four covalent bonds</u> in a <u>very rigid giant covalent structure</u>. This structure is why diamond is very, <u>very hard</u>. It contains only covalent bonds, which are very strong.

Graphite

Each carbon atom only forms <u>three covalent bonds</u>, creating <u>layers</u> which are free to <u>slide over each other</u>, and leaving <u>free electrons</u>, so graphite is the only <u>non-metal</u> which <u>conducts electricity</u>.

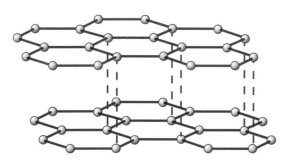

Silicon Dioxide

Sometimes called <u>silica</u>, this is what <u>sand</u> is made of.

<u>Each grain of sand</u> is <u>one giant structure</u> of silicon and oxygen.

Giant Covalent Structures can earn you Big Marks

There are two types of covalently bonded substances — and they're totally different. Make sure you know all the details about them, and the examples too. This is just easy marks to be won... or lost.

Metallic Structures

Metal Properties are all due to the *Sea of Free Electrons*

1) <u>Metals</u> also consist of a <u>giant structure</u>.

2) <u>Metallic bonds</u> involve the all-important "<u>free electrons</u>", which produce <u>all</u> the properties of metals. These free electrons come from the <u>outer shell</u> of <u>every</u> metal atom in the structure.

3) These electrons are <u>free to move</u> and so metals <u>conduct heat and electricity</u>.

4) These electrons also <u>hold the atoms together</u> in a regular structure.

5) They also allow the atoms to <u>slide over each</u> other causing metals to be <u>malleable</u> (which means you can do useful things with them, like roll them into sheets).

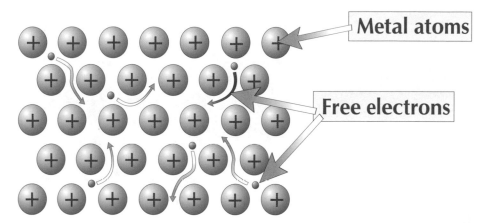

Metal atoms

Free electrons

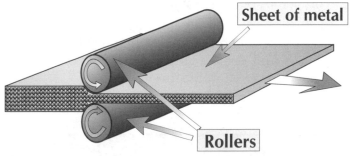

Sheet of metal

Rollers

Identifying the bonding in a substance *by its properties*

If you've learnt the properties of the <u>four types</u> of substance properly, together with their <u>names</u> of course, then you should be able to easily <u>identify</u> most substances just by the way they <u>behave</u> as either: <u>ionic</u>, <u>giant-covalent</u>, <u>simple molecular covalent</u>, or <u>metallic</u>.

The way they're likely to test you in the Exam is by describing the <u>physical properties</u> of a substance and asking you to decide <u>which type of bonding</u> it has and therefore what type of material it is. If you know your bonding you'll have no trouble at all. If not, you're going to struggle.

Properties are important
Think about the tell-tale signs for each type of substance, e.g., if something has a low boiling point, and doesn't conduct electricity, then it's going to be a simple molecular substance. If it's malleable...

Elements, Compounds and Mixtures

You'd better be sure you know the subtle differences between these.

Elements consist of one type of atom only

Quite a lot of everyday substances are elements:

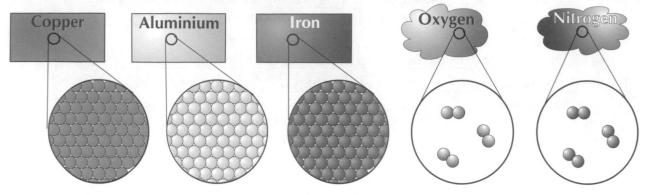

Copper   Aluminium   Iron   Oxygen   Nitrogen

Mixtures are easily separated

1) Air is a mixture of gases. The oxygen, nitrogen, argon and carbon dioxide can all be separated out quite easily.

2) There is no chemical bond between the different parts of a mixture.

3) The properties of a mixture are just a mixture of the properties of the separate parts.

4) A mixture of iron powder and sulphur powder will show the properties of both iron and sulphur. It will contain grey magnetic bits of iron and bright yellow bits of sulphur.

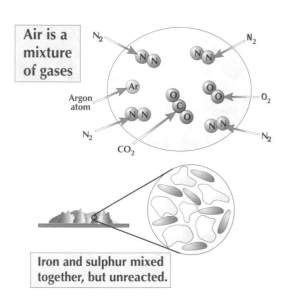

Air is a mixture of gases

N_2 N_2 Argon atom N_2 CO_2 N_2 O_2

Iron and sulphur mixed together, but unreacted.

Compounds are chemically bonded

1) Carbon dioxide is a compound formed from a chemical reaction between carbon and oxygen.

2) It's very difficult to separate the two original elements out again.

3) The properties of a compound are totally different from the properties of the original elements.

4) If iron and sulphur react to form iron sulphide, the compound formed is a grey solid lump, and doesn't behave anything like either iron or sulphur.

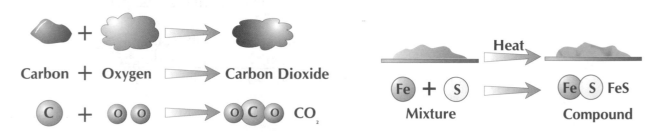

Carbon + Oxygen ⟶ Carbon Dioxide

C + O O ⟶ O C O CO_2

Heat

Fe + S ⟶ Fe S FeS
Mixture   Compound

Don't mix these up

Elements, mixtures and compounds. To most people they sound like basically the same thing. But not to GCSE Examiners. You make mighty sure you remember the differences between them.

Common Tests and Hazard Symbols

You need to know the <u>five easy lab tests</u> on the next two pages:

1) Chlorine **bleaches damp litmus paper**

Chlorine turns damp litmus paper <u>white</u>.

2) Oxygen **relights a glowing splint**

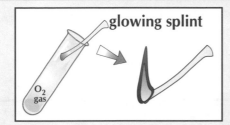

The standard test for <u>oxygen</u> is that it <u>relights a glowing splint</u>.

3) Carbon dioxide **turns limewater milky**

<u>Carbon dioxide</u> can be detected by bubbling it through <u>limewater</u>. It turns the limewater <u>cloudy</u>.

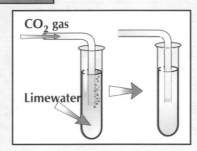

4) The **three lab tests for Water**

Water can be detected in three ways:

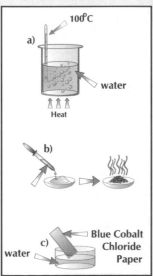

a) by its <u>boiling point</u> of <u>100°C</u>

b) by turning <u>white anhydrous copper sulphate</u> to <u>blue hydrated copper sulphate</u> (and getting hot)

c) by turning <u>anhydrous cobalt chloride paper</u> from <u>blue</u> to <u>pink</u>.

There's one more to go on the next page.

How else are you going to tell your colourless gases apart?

It's no good just letting your eyes drift lazily across the page and thinking "I know all that stuff". You've got to really make sure you learn it because the Examiners love to ask questions on this.

Common Tests and Hazard Symbols

Don't forget Hydrogen.

5) Lab test for Hydrogen — *the notorious "Squeaky pop"*

Just bring a <u>lighted splint</u> near the gas with air around. If it's hydrogen it'll <u>burn with a ("squeaky") pop</u> as it reacts with the oxygen in the air to form H_2O.

Learn these hazard symbols and you might be able to pick up some easy marks.

Hazard Symbols

Oxidising
<u>Provides oxygen</u> which allows other materials to <u>burn more fiercely</u>.
<u>Example</u>: Liquid oxygen.

Harmful
Similar to toxic but <u>not quite as dangerous</u>.
<u>Example</u>: Petrol, meths.

Highly Flammable
<u>Catches fire</u> easily.
<u>Example</u>: Petrol.

Irritant
Not corrosive but <u>can cause reddening or blistering of the skin</u>.
<u>Examples</u>: Bleach, etc.

Toxic
<u>Can cause death</u> either by swallowing, breathing in, or absorption through the skin.
<u>Example</u>: Cyanide.

Corrosive
<u>Attacks and destroys living tissues</u>, including eyes and skin.
<u>Example</u>: Sulphuric acid, sodium hydroxide.

<u>Note</u>: *The official hazard symbol for "harmful" and "irritant" is a black cross. Some products add an "h" or "i" to show the difference.*

Better to be safe than sorry
You probably won't be lucky enough to get a question on hazard symbols in the Exam, but then it's best to learn them anyway so that you don't throw away easy marks. Besides, they're jolly useful to know.

Warm-Up and Worked Exam Questions

It's time for more questions. Try these straightforward ones first.

Warm-up Questions

1) Write down the symbols, with charge signs, for the ions of the following elements: lithium, sodium, potassium, beryllium, magnesium, calcium, oxygen, fluorine and chlorine.
2) Water and diamond are both covalent compounds. Explain why water melts at 0 °C and diamond at more than 3000 °C.
3) Common salt (sodium chloride) and aluminium both have giant structures. Explain why aluminium conducts electricity and solid salt does not.
4) Describe a test for carbon dioxide.
5) What does the hazard symbol of 'skull and crossbones' on a container tell us about its contents?

Worked Exam Question

Structures are really important — you need to know how they affect a substance's properties. Work through this example, then have a go at the questions that follow.

1   Magnesium and oxygen react to form magnesium oxide. The symbol equation is shown below.

$$2Mg_{(s)} + O_{2(g)} \rightarrow 2MgO_{(s)}$$

a) Explain why magnesium is solid at room temperature.

    The 'free electrons' produce strong metallic bonds which hold the atoms in a giant structure. The bonds are hard to break, so the melting point is high.

    (3 marks)

b) Explain why oxygen is a gas at room temperature.

    Oxygen atoms form small covalent molecules, and there are only very weak intermolecular forces between the oxygen molecules. These bonds are very easily broken, so the boiling point is low.

    (3 marks)

c) What type of bonding exists in magnesium oxide?

    Ionic   *Between Mg^{2+} ions and O^{2-} ions.*

    (1 mark)

Exam Questions

1   Name a gas that:

a)   bleaches damp litmus paper

    ....
    (1 mark)

b)   relights a glowing splint

    ....
    (1 mark)

c)   turns limewater milky

    ....
    (1 mark)

d)   is a compound made of simple molecules

    ....
    (1 mark)

2   A chemist finds a beaker in the lab containing a clear liquid. She believes it is water, but is not sure.

a)   Suggest two safe methods of confirming that the liquid is water.

    1...

    ....

    2...

    ....
    (2 marks)

b)   Water is a covalent compound. Name two properties of water that suggest it has a simple molecular structure.

    ....

    ....
    (2 marks)

c)   The chemist confirms that the liquid in the beaker is water.
    She accidentally drops some sodium chloride into the beaker. Will the resulting solution conduct electricity? Explain your answer.

    ....

    ....

    ....
    (3 marks)

Exam Questions

3   Four unknown substances are labelled A, B, C and D.

	Boiling point	Conducts electricity?	Solubility in water
A	High	Yes when molten or in solution	Dissolves readily
B	High	Yes	Does not dissolve
C	Very High	No	Does not dissolve
D	Low	No	Does not dissolve

a)   Study the table above showing the properties of each substance.

   i)   Which substance is a metal?

   ....

   (1 mark)

   ii)   Explain why metals are good conductors of electricity.

   ....

   ....

   ....

   (3 marks)

b)   You are told that one of the substances might be hydrogen.

   i)   Which substance is most likely to be hydrogen?

   ....

   (1 mark)

   ii)   Describe a test you could do to confirm that this substance is hydrogen.

   ....

   ....

   (2 marks)

c)   i)   What type of bonding is present in substance A?

   ....

   (1 mark)

   ii)   Explain why substance A has a high boiling point.

   ....

   ....

   ....

   (2 marks)

Revision Summary

These certainly aren't the easiest questions you're going to come across. That's because they test what you know without giving you any clues. At first you might think they're impossibly difficult. Eventually you'll realise that they simply test whether you've learnt the stuff or not.
If you're struggling to answer these then you need to do some serious learning.

1) What are the three states of matter?
2) Describe the bonding and atom spacing in all three states.
3) Describe the physical properties of each of these three states of matter.
4) What are the ways of changing between the three states of matter?
5) Explain what goes on in all three processes, in terms of bonds and heat energy.
6) Sketch a heating graph and a cooling graph, with lots of labels.
7) Explain why these graphs have flat spots.
8) Sketch an atom. Give five details about the nucleus and five details about the electrons.
9) What are the three particles found in an atom?
10) Do a table showing their relative masses and charges.
11) How do the numbers of these particles compare to each other in a neutral atom?
12) What do the mass number and atomic number represent?
13) Explain what an isotope is. Give a well-known example.
14) List five facts (or "Rules") about electron shells.
15) What is ionic bonding? Which kind of atoms like to do ionic bonding?
16) Why do atoms want to form ionic bonds?
17) What is covalent bonding?
18) Why do some atoms do covalent bonding instead of ionic bonding?
19) Give five examples of covalent molecules, and sketch diagrams showing the electrons.
20) What kind of ions are formed by elements in Groups I and II, and those in Groups VI and VII?
21) Draw a diagram of a giant ionic lattice and give three features of giant ionic structures.
22) List the three main properties of ionic compounds.
23) What are the two types of covalent substances? Give three examples of each type.
24) Give three physical properties for each of the two types of covalent substance.
25) Explain how the bonding in each type of covalent substance causes its physical properties.
26) What is special about the bonding in metals?
27) List the three main properties of metals and explain how the metallic bonding causes them.
28) What is the difference between elements, mixtures and compounds?
29) Give three examples each of elements, mixtures and compounds.
30) Give full details of the lab tests for:
      Chlorine, oxygen, carbon dioxide, water (3), hydrogen.
31) Sketch the six Hazard Symbols, explain what they mean, and give an example for each.

Fractional Distillation of Crude Oil

1) Crude oil is formed from the buried remains of plants and animals — it's a fossil fuel. Over millions of years, with high temperature and pressure, the remains turn to crude oil which can be drilled up.

2) Crude oil is a mixture of hydrocarbons of different sized molecules.

3) Hydrocarbons are basically fuels such as petrol and diesel. They're made of just carbon and hydrogen.

4) The bigger and longer the molecules, the less runny the hydrocarbon (fuel) is.

5) Fractional distillation splits crude oil up into its separate fractions.

6) The shorter the molecules, the lower the temperature at which that fraction condenses.

Crude Oil is *Split* into *Separate Hydrocarbons* (fuels)

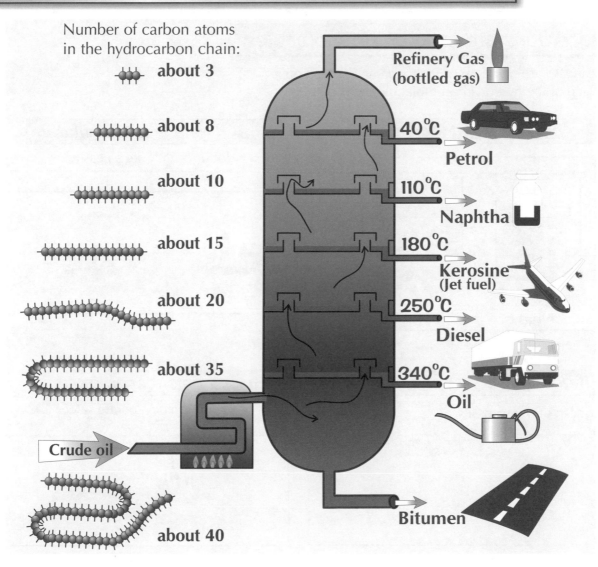

Number of carbon atoms in the hydrocarbon chain:

about 3

about 8

about 10

about 15

about 20

about 35

Crude oil

about 40

Refinery Gas (bottled gas)

40°C — Petrol

110°C — Naphtha

180°C — Kerosine (Jet fuel)

250°C — Diesel

340°C — Oil

Bitumen

The fractionating column works continuously, with heated crude oil vapour piped in at the bottom and the various fractions being constantly tapped off at the different levels where they condense.

Learn it all — you know the drill

In the Exam you might be asked about the uses of hydrocarbons, or from what part of a fractionating column you'd expect to get petrol, or about any of the other tiny little details — so learn them all.

Hydrocarbons

Crude oil is a very big part of *modern life*

1) There's a <u>massive industry</u> with scientists working to find oil reserves, take it out of the ground, and turn it into useful products.

2) It provides the <u>fuel</u> for most modern transport.

3) It also provides the <u>raw materials</u> for making various <u>chemicals</u> including <u>plastics</u>. The world without plastics? Why, it would be the end of civilisation as we know it...

4) Oil can be seriously bad news for the <u>environment</u>. Oil slicks at sea, old engine oil down the drain, plastics that won't rot if you throw them away... <u>Some</u> things can be <u>recycled</u> though, which helps.

Hydrocarbons are *long chain molecules*

Hydrocarbons are made up of hydrogen and carbon only.
As the <u>size</u> of the <u>hydrocarbon molecule increases</u>:

1) The boiling point increases

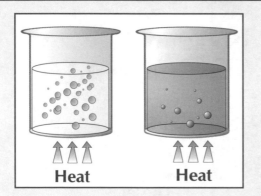

Heat   **Heat**

2) It gets less flammable

(doesn't catch fire so easily)

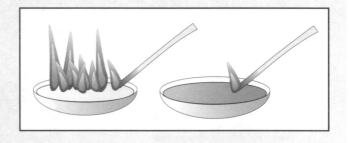

3) It gets more viscous

(doesn't flow so easily)

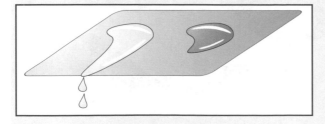

4) It gets less volatile

(i.e. doesn't evaporate so easily)

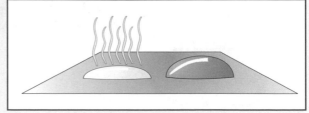

The <u>vapours</u> of the more <u>volatile</u> hydrocarbons are <u>very flammable</u> and pose a serious <u>fire risk</u>. So don't smoke at the petrol station.

Length is important

As the length of the hydrocarbon chain increases, the properties alter, making different hydrocarbons suitable for different things. You'd be in trouble if you tried to put bitumen in your petrol tank.

Combustion of Hydrocarbons

Combustion of Hydrocarbons

Hydrocarbons are often used as <u>fuels</u> because they <u>burn well</u>. Here's what happens when you burn hydrocarbons.

Complete combustion of Hydrocarbons is safe

The <u>complete combustion</u> of any hydrocarbon in oxygen will produce only <u>carbon dioxide</u> and <u>water</u> as waste products, which are both quite <u>clean</u> and <u>non-poisonous</u>.

> **hydrocarbon + oxygen → carbon dioxide + water** (+ energy)

Many <u>gas room heaters</u> release these <u>waste gases</u> into the room, which is perfectly fine. As long as the gas heater is working properly and the room is well ventilated there's no problem. When there's <u>plenty of oxygen</u> the gas burns with a <u>clean blue flame</u>.

But Incomplete combustion of Hydrocarbons is NOT safe

If there <u>isn't enough oxygen</u> the combustion will be <u>incomplete</u>.
This gives <u>carbon monoxide</u> and <u>carbon</u> as waste products.

> **hydrocarbon + oxygen → CO_2 + H_2O + carbon monoxide + carbon** (+ energy)

The <u>carbon monoxide</u> is a <u>colourless</u>, <u>odourless</u> and <u>poisonous</u> gas and it's <u>very dangerous</u>. Every year people are <u>killed</u> while they sleep due to <u>faulty</u> gas fires and boilers filling the room with <u>deadly carbon monoxide</u>, CO, and nobody realising. The black carbon given off produces <u>sooty marks</u> and is a <u>clue</u> that the fuel is <u>not</u> burning fully. The smoky yellow flame is also an indicator of incomplete combustion.

The one burning question is... have you learnt it all...

The details for complete and incomplete combustion are really important, and <u>worth lots of marks in the Exam</u>. Once you've learnt them, cover the page and write the equations out from memory.

Cracking Hydrocarbons

Cracking — *splitting up* long chain hydrocarbons

1) Long chain hydrocarbons form thick gloopy liquids like tar which aren't all that useful.
2) The process called cracking turns them into shorter molecules which are much more useful.

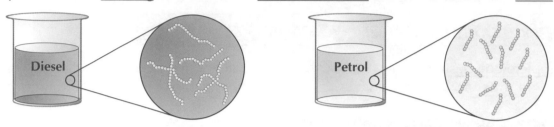

3) Cracking is a form of thermal decomposition, which just means breaking molecules down into simpler molecules by heating them.
4) A lot of the longer molecules produced from fractional distillation are cracked into smaller ones because there's more demand for products like petrol and paraffin (jet fuel) than for diesel or lubricating oil.
5) More importantly, cracking produces extra alkenes which are needed for making plastics.

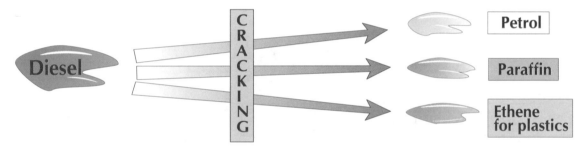

Industrial Conditions for Cracking: *hot, plus a catalyst*

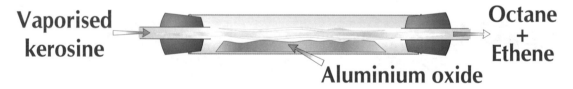

1) Vaporised hydrocarbons are passed over powdered catalyst at about 400°C – 700°C.
2) Aluminium oxide is the catalyst used.
3) The long chain molecules split apart or "crack" on the surface of the bits of catalyst.

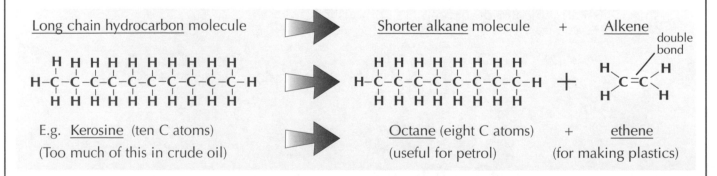

Long chain hydrocarbon molecule ➡ Shorter alkane molecule + Alkene

E.g. Kerosine (ten C atoms) → Octane (eight C atoms) + ethene
(Too much of this in crude oil)   (useful for petrol)   (for making plastics)

Short chain hydrocarbons are more useful as fuels

Five details about the whys and wherefores, three details of the industrial conditions and a specific example showing typical products: a shorter chain alkane and an alkene. LEARN IT ALL.

Warm-Up and Worked Exam Questions

Fractional distillation and cracking hydrocarbons are not particularly hard topics. If you get a question that asks for a balanced equation, make sure that you balance that equation, or you'll lose marks.

Warm-up Questions

1) What elements are hydrocarbons made from?
2) What is the name of the process used to separate crude oil into products like petrol?
3) Write down the two products formed by the complete combustion of methane (a hydrocarbon).
4) What is the name of the process used to turn long chain hydrocarbons into shorter molecules?

Worked Exam Question

Take a look at this worked exam question. It's not too hard but it should give you a good idea of what to write.

1   Long chain hydrocarbons are cracked to turn them into shorter molecules.

a)   Why are the longer chain hydrocarbons produced by fractional distillation cracked into smaller molecules?

There is more demand for the shorter molecules than the longer

molecules, since they are much more useful.

(1 mark)

b)   Cracking octane (C_8H_{18}) can produce two hydrocarbons, each containing four carbon atoms.

i)   Write a balanced equation for this cracking reaction.

$$C_8H_{18} \rightarrow C_4H_{10} + C_4H_8$$

(2 marks)

ii)   Suggest two conditions used for this cracking reaction.

Heat (400°C - 700°C)

and a catalyst (aluminium oxide)

You need two conditions to get the two marks.

(2 marks)

c)   Cracking hydrocarbons produces alkanes and alkenes.

i)   What materials are alkenes used to make?

Plastics

(1 mark)

ii)   What feature of the structure of an alkene allows them to be used for this?

A double bond

(1 mark)

Exam Questions

1   Crude oil is separated into fuels such as petrol, kerosine and diesel by fractional distillation.
     a)   Look at this diagram of a fractionating column.

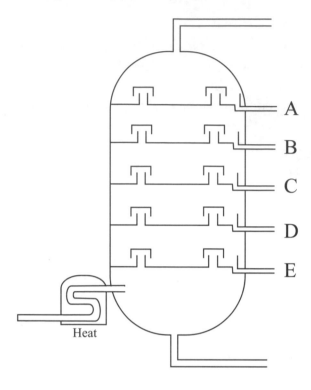

Heat

     i)   Petrol has a boiling range of 40-100°C. Write down the letter from the diagram
          that shows where petrol would be piped off.

          ....

          (1 mark)

     ii) Petrol has a lower boiling point than kerosene. Write down which of the two
          fuels has more carbon atoms in its molecules. Explain your answer by stating
          the relationship between the boiling range and the number of carbon atoms in
          the molecule.

          ....

          ....

          (2 marks)

     b)   The products of the fractional distillation of crude oil are often used as fuels.
          Name another use of one of these products.

          ....

          (1 mark)

Exam Questions

2   Alkanes such as methane and ethane are burnt as fuels to produce energy.

a)   Write down the two products of the complete combustion of ethane.

....

(2 marks)

b)   The complete combustion of ethane produces a blue flame.

i)   Describe the flame produced by the incomplete combustion of ethane.

....

....

(2 marks)

ii)   Write a word equation for the incomplete combustion of ethane.

....

....

(2 marks)

c)   Some room heaters use alkanes such as methane as a fuel. These gas room heaters need to be checked regularly to make sure they are safe. Every year people die from breathing in the fumes from heaters which have not had safety checks. Explain how the heaters produce these fumes, and why they can kill people.

Credit will be given for the use of correct terms in your answer, and also for the correct use of spelling, punctuation and grammar.

....

....

....

....

(3 marks)

Alkanes and Alkenes

Crude oil contains both <u>alkanes</u> and <u>alkenes</u>. Know the differences between them.

ALKANES have all C–C SINGLE bonds

1) They're made up of <u>chains</u> of carbon atoms with <u>single covalent bonds</u> between them.

2) They're called <u>saturated hydrocarbons</u> because they have <u>no spare bonds</u> left.

3) This is also why they <u>don't</u> decolourise <u>bromine water</u> — no spare bonds.

4) They <u>won't</u> form <u>polymers</u> — same reason again, <u>no spare bonds</u>.

5) The first four alkanes are <u>methane</u> (natural gas), <u>ethane</u>, <u>propane</u> and <u>butane</u>.

6) They <u>burn cleanly</u> producing <u>carbon dioxide</u> and <u>water</u>.

Bromine water with alkane — still brown.

1) Methane

Formula: CH_4

$$\begin{array}{c} H \\ | \\ H-C-H \\ | \\ H \end{array}$$

(natural gas)

2) Ethane

Formula: C_2H_6

$$\begin{array}{cc} H & H \\ | & | \\ H-C-C-H \\ | & | \\ H & H \end{array}$$

3) Propane

Formula: C_3H_8

$$\begin{array}{ccc} H & H & H \\ | & | & | \\ H-C-C-C-H \\ | & | & | \\ H & H & H \end{array}$$

4) Butane

Formula: C_4H_{10}

$$\begin{array}{cccc} H & H & H & H \\ | & | & | & | \\ H-C-C-C-C-H \\ | & | & | & | \\ H & H & H & H \end{array}$$

Alkane anybody who doesn't learn this lot properly

Drawing alkane molecules is as easy as falling off a log. Just write the correct number of Cs in a row, join them together, and add enough H atoms so that all the Cs are attached to four other atoms. Easy.

Alkanes and Alkenes

ALKENES have a C=C DOUBLE bond

1) They're <u>chains</u> of carbon atoms with one <u>double bond</u>.

2) They are called <u>unsaturated hydrocarbons</u> because they have one <u>spare bond</u> left.

3) This is why they <u>will</u> decolourise <u>bromine water</u>. They form <u>bonds</u> with bromide ions.

4) They form <u>polymers</u> by <u>opening up</u> their double bonds to "<u>hold hands</u>" in a long chain.

5) The first three alkenes are <u>ethene</u>, <u>propene</u> and <u>butene</u>.

6) They tend to burn with a <u>smoky flame</u>, producing <u>soot</u> (carbon).

Bromine water with alkene — decolourised

1) Ethene

Formula: C_2H_4

2) Propene

Formula: C_3H_6

3) Butene

Formula: C_4H_8

Important Notes

1) <u>Bromine water</u> is the <u>standard test</u> to distinguish between alkanes and alkenes.

2) <u>Alkenes</u> are <u>more reactive</u> due to the <u>double bond</u> all poised and ready to just <u>pop open</u>.

3) Notice the <u>names</u>: "<u>Meth-</u>" means "<u>one</u> carbon atom", "<u>eth-</u>" means "<u>two</u> C atoms", "<u>prop-</u>" means "<u>three</u> C atoms", "<u>but-</u>" means "<u>four</u> C atoms", etc. The only difference then between the names of <u>alkanes</u> and <u>alkenes</u> is just the "<u>-ane</u>" or "<u>-ene</u>" on the end.

4) <u>All alkanes</u> have the formula: C_nH_{2n+2}. <u>All alkenes</u> have the formula: C_nH_{2n}.

Not at Alkene to revise? Tough, you need to know this

Drawing alkenes is only slightly more complicated than alkanes. Just remember that each C atom needs to make 4 bonds, then add the right number of Hs, paying attention to where the double bond is.

Polymers and Plastics

Polymers and plastics were first discovered in about 1933. By 1970 it was all too late. Leather seats and lovely wooden dashboards were replaced by cheap, wipe-clean plastic.

Alkenes open their **double bonds** to form **Polymers**

Under a bit of <u>pressure</u> and with a <u>catalyst</u> to help it along, many <u>small alkenes</u> will open up their <u>double bonds</u> and "join hands" to form <u>very long chains</u> called <u>polymers</u>. This process is called <u>polymerisation</u>.
<u>Ethene</u> becoming polyethene or "polythene" is the easiest example:

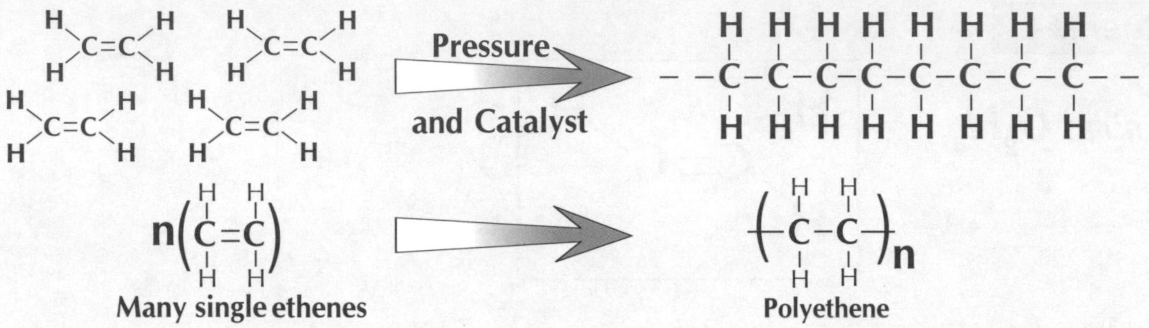

Many single ethenes → **Polyethene**

Other **small alkenes** do a similar trick:

<u>Propene</u> can form <u>polypropene</u>:

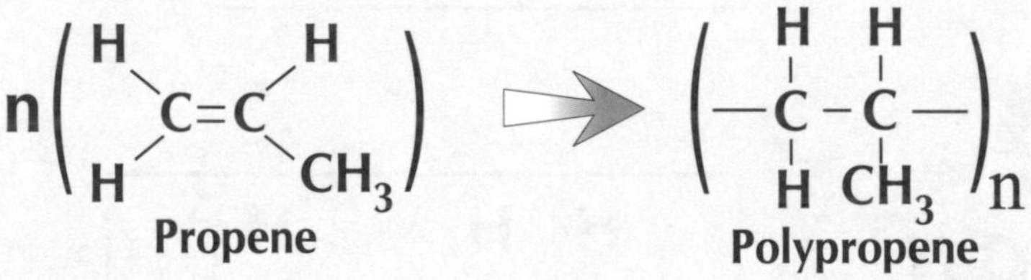

Propene → **Polypropene**

A molecule called <u>styrene</u> will <u>polymerise</u> into <u>polystyrene</u>:

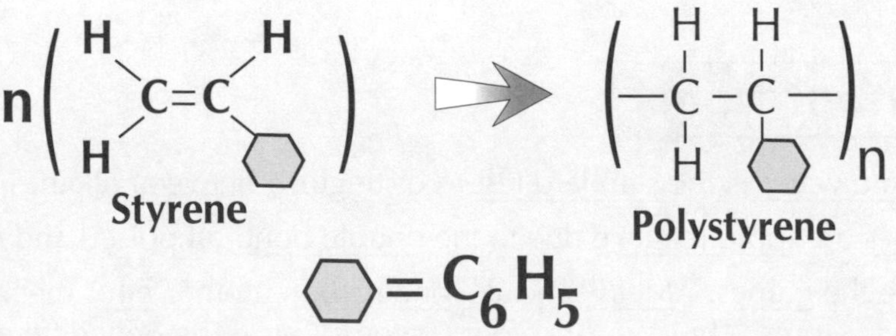

Styrene → **Polystyrene**

$$\bigcirc = C_6H_5$$

Breaking the double bond lets alkenes join together

Learn what polymerisation is and practise the diagrams for ethene. It's really important that you understand how polymerisation works. Now cover the page and write down what you know.

Polymers and Plastics

There are loads of *Plastics* with *loads* of different uses

1) Polythene

1) Very <u>cheap</u> and <u>strong</u>
2) Easily <u>moulded</u>

Plastic bags

Bottles

Buckets

Bowls

2) Polypropene

1) Forms <u>strong fibres</u>
2) Has <u>high elasticity</u>

Crates

Ropes

Carpet

3) Polystyrene

1) <u>Cheap</u> and easily <u>moulded</u>
2) Can be <u>expanded</u> into <u>foam</u>

Radio outer cases

Foam Packaging

4) PVC

1) <u>Cheap</u> and easily <u>moulded</u>
2) Can be <u>expanded</u> into <u>foam</u>

Plastic sheets

Electric wire insulation

Records

Most plastics *don't rot*, so they're hard to get rid of

1) Most plastics aren't '<u>biodegradable</u>' — they're not broken down by microorganisms, so they <u>don't rot</u>.
2) It's difficult to get rid of them — if you bury them in a landfill site, they'll still be there <u>years later</u>. The best thing is to <u>recycle</u> them if you can.

Fantastic Plastic can get you easy marks
Aren't plastics great. Learn all the great stuff you can make with the different types of plastics so you'll have lots of great examples for the Exam. Nothing hard here, just a few great facts to learn. Great!

Warm-Up and Worked Exam Questions

The hardest thing you'll probably be asked to do with alkenes is to draw out a graphical formula for part of a polymer. The rest is fairly simple.

Warm-up Questions

1) What kind of bonds do alkanes have?
2) What is a saturated hydrocarbon?
3) What are hydrocarbons with one C=C double bond called?
4) Can alkanes form polymers?
5) Can alkenes form polymers?

Worked Exam Question

This worked example is about the formulae of alkenes, and the standard lab test for alkenes. It's a topic that's likely to appear in the real Exam.

1   Crude oil contains saturated and unsaturated hydrocarbons.

(a) What is the name of this hydrocarbon?

$$H-\overset{\overset{\displaystyle H}{|}}{\underset{\underset{\displaystyle H}{|}}{C}}-\overset{\overset{\displaystyle H}{|}}{\underset{\underset{\displaystyle H}{|}}{C}}=C\overset{\diagup H}{\diagdown H}$$

There are three carbon atoms, so it's prop-something, and a double bond, so it's something -ene.

.... *Propene* ....

(1 mark)

(b) Describe the laboratory test for an unsaturated hydrocarbon.

The hydrocarbon is bubbled through bromine water. If the

hydrocarbon is unsaturated, the bromine water goes colourless.

If it is saturated, the bromine water stays brown

....

Don't leave out the fact that bromine water is brown to start with and goes colourless when there's a double bond about.

(2 marks)

(c) Explain why unsaturated hydrocarbons are more reactive than saturated hydrocarbons.

Unsaturated hydrocarbons have spare bonds. The double bonds

can open and form new bonds with other atoms. Saturated

hydrocarbons have no spare bonds.

(2 marks)

You need to say that saturated hydrocarbons don't have spare bonds.

Exam Questions

1   Alkenes are the raw materials for plastics.

   (a) What industrial process is a major source of alkenes?

   ....
   (1 mark)

   (b) This is a graphical formula for propene.

$$H-\overset{\displaystyle H}{\underset{\displaystyle H}{C}}-\overset{\displaystyle}{\underset{\displaystyle H}{C}}=C\overset{\displaystyle H}{\underset{\displaystyle H}{<}}$$

      (i)   Draw a graphical formula of the repeating unit in polypropene.

      (2 marks)

      (ii) What conditions are required for propene to polymerise?

      ....
      (2 marks)

      (iii) Polypropene is used to make crates and fibres for carpet. Suggest a reason why it is
            suitable for these uses.

      ....
      (2 marks)

   (c) Polythene is used to make plastic bags and bottles. It does not biodegrade.

      (i)   Polythene can be disposed of by being buried. Suggest a disadvantage of disposing
            of polythene products in this way.

      ....
      (1 mark)

      (ii) Polythene can be recycled. Suggest one reason for recycling more polythene (other
            than the difficulties of disposing of it).

      ....
      (1 mark)

      (iii) Suggest a reason why recycling or reusing polythene products might be unpopular.

      ....
      (1 mark)

Metal Ores from the Ground

Rocks, Minerals and Ores

1) A rock is a mixture of minerals.

2) A mineral is any solid element or compound found naturally in the Earth's crust.
   Examples: Diamond (carbon), quartz (silicon dioxide), bauxite (Al_2O_3).

3) A metal ore is defined as a mineral or minerals which contain enough metal in them
   to make it worthwhile extracting the metal from it.

4) There's a limited amount of minerals and ores — they're "finite resources".

Metals are **extracted** from ores using **Carbon** or **Electrolysis**

1) Extracting a metal from its ore involves a chemical reaction to separate the metal out.

2) In many cases the metal is found as an oxide. There are three ores you need to know:

> a) **Iron ore** is called **Haematite**, which is iron(III) oxide.
>   Formula Fe_2O_3.
>
> b) **Aluminium ore** is called **Bauxite**, which is aluminium oxide.
>   Formula Al_2O_3.
>
> c) **Copper ore** is called **Malachite**, which is copper(II) carbonate.
>   Formula $CuCO_3$.

3) The two common ways of extracting a metal from its ore are:

   a) Chemical reduction using carbon or carbon monoxide
   b) Electrolysis

   Iron is extracted by chemical reduction
   Aluminium is extracted by electrolysis

4) Gold is one of the few metals found as a metal
   rather than in a chemical compound (an ore).
   That's why people find pure gold nuggets...

Memorise the three ores

This page has all the basics you need to know about ores before you go on to tackle the harder stuff on
the next few pages. Make sure you know the three ores, including their formulae.
You need to practise repeating the details from memory. That's the only effective method.

Metal Ores from the Ground

More Reactive Metals are **Harder to Get**

1) The <u>more reactive</u> metals took <u>longer</u> to be discovered. (e.g. aluminium, sodium)

2) The <u>more reactive</u> metals are also <u>harder to extract</u> from their mineral ores.

3) The above <u>two facts</u> are obviously <u>related</u>. It's <u>obvious</u> when you think about it...

> *Even primitive people could find gold easily enough just by scrabbling about in streams, and then melt it into ingots and jewellery and statues. But coming up with a fully operational electrolysis plant to extract sodium metal from rock salt, just by paddling about a bit... unlikely.*

The Position of **Carbon** in the Reactivity Series decides it...

1) Metals <u>higher than carbon</u> in the reactivity series have to be extracted using <u>electrolysis</u>.

2) Metals <u>below carbon</u> in the reactivity series can be extracted by <u>reduction</u> using <u>carbon</u>.

3) This is obviously because carbon <u>can only take the oxygen away from</u> (reduce) metals which are <u>less reactive</u> than carbon <u>itself</u> is.

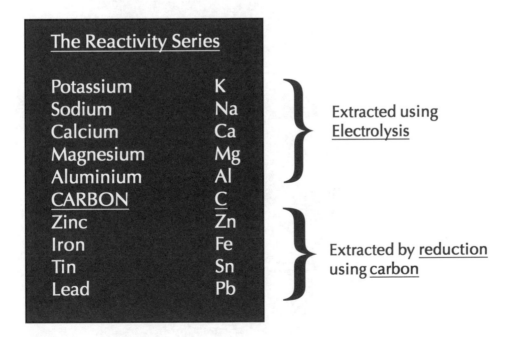

The Reactivity Series

Potassium	K	
Sodium	Na	Extracted using
Calcium	Ca	<u>Electrolysis</u>
Magnesium	Mg	
Aluminium	Al	
<u>CARBON</u>	<u>C</u>	
Zinc	Zn	
Iron	Fe	Extracted by <u>reduction</u>
Tin	Sn	using <u>carbon</u>
Lead	Pb	

Reactivity dictates how metals are extracted

Make sure you learn the reactivity series above. Try writing each element on a scrap of paper and then arranging them in the correct order from memory. If you know this, you'll know which metals can be reduced using carbon, and which need electrolysis.

Extracting Iron — the Blast Furnace

Iron is a very common element in the Earth's crust, but good iron ores are only found in a few select places around the world, such as Australia and Canada. Iron is extracted from haematite, Fe_2O_3, by reduction (i.e. removal of oxygen) in a blast furnace.

The Raw Materials are **Iron Ore**, **Coke** and **Limestone**

Iron ore contains the iron — which is pretty important.

Coke is almost pure carbon. This is for reducing the iron oxide to iron metal.

Limestone takes away impurities in the form of slag.

You really do need to know all these details about what goes on in a blast furnace.

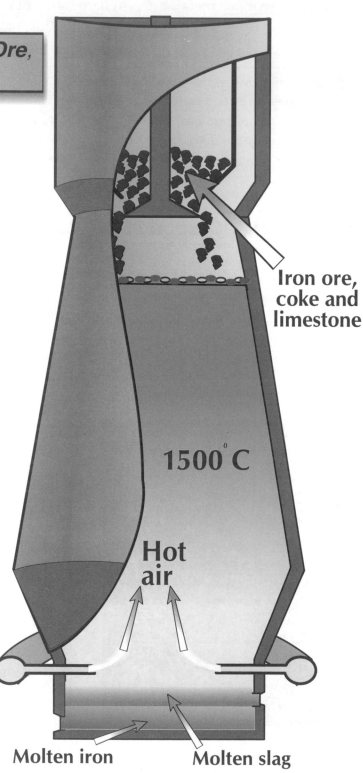

Iron ore, coke and limestone

1500°C

Hot air

Molten iron

Molten slag

Learn the facts about Iron Extraction
Look at the diagram, learn it, cover it up and then sketch it from memory in as much detail as you can manage. When you can do that really well, go on to the next page and lap up those equations.

Extracting Iron — the Blast Furnace

Reducing the **Iron Ore** to **Iron**:

1) Hot air is blasted into the furnace making the coke burn much faster than normal and the temperature rises to about 1500°C.

2) The coke burns and produces carbon dioxide:

$$C \;+\; O_2 \;\rightarrow\; CO_2$$
carbon +   oxygen   →   carbon dioxide

3) The CO_2 then reacts with unburnt coke to form CO:

$$CO_2 \;+\; C \;\rightarrow\; 2CO$$
carbon dioxide +   carbon →   carbon monoxide

4) The carbon monoxide then reduces the iron ore to iron:

$$3CO \;+\; Fe_2O_3 \;\rightarrow\; 3CO_2 \;+\; 2Fe$$
carbon monoxide + iron(III)oxide → carbon dioxide + iron

5) The iron is of course molten at this temperature and it's also very dense so it runs straight to the bottom of the furnace where it's tapped off.

Removing the **Impurities**:

1) The main impurity is sand (silicon dioxide). This is still solid even at 1500°C and would tend to stay mixed in with the iron. The limestone removes it.

2) The limestone is decomposed by the heat into calcium oxide and CO_2.

$$CaCO_3 \;\rightarrow\; CaO + CO_2$$

3) The calcium oxide then reacts with the sand to form calcium silicate, or slag, which is molten and can be tapped off:

$$CaO + SiO_2 \;\rightarrow\; CaSiO_3 \text{ (molten slag)}$$

4) The cooled slag is solid, and is used for:
   1) Road building   2) Fertiliser

You'll need these equations in the Exam

Equations, equations, too many equations. Well, you're going to have to learn them. The big nasty one for the actual reduction is the most important, but you'll need the others too.

Warm-Up and Worked Exam Questions

There are quite a few equations to remember here, which is where most people trip up.

Warm-up Questions

1) What is an ore?
2) What's the usual method for getting iron out of iron ore?
3) Bauxite is an ore of which metal?
4) Is gold found in the ground as an ore or as the metal?

Worked Exam Question

Take a look at this worked example.

1   In a blast furnace, coke is burnt to reduce iron ore to iron.

a)   What is the main element in coke?

   Carbon

   (1 mark)

b)   Coke burns to produce carbon dioxide.

   i)   Write a balanced equation for the burning of coke.

   $C + O_2 \rightarrow CO_2$

   (2 marks)

   ii)   Carbon dioxide reacts with unburnt coke. Complete this equation to show this.

   $CO_2 + C \rightarrow$ *2CO*   Remember that all-important '2' — you
                                         need it to balance the equation.

   (1 mark)

c)   Carbon monoxide (CO) reacts with iron(III) oxide (Fe_2O_3) to produce iron.

   i)   Write a balanced equation to show this.

   $3CO + Fe_2O_3 \rightarrow 2Fe + 3CO_2$

   (2 marks)

   ii)   What happens to the iron produced in this reaction?

   It collects at the bottom of the furnace and is

   tapped off.

   (1 mark)

Exam Questions

1   (a)   This table shows the reactivity series.

Potassium   More reactive
Sodium
Calcium
Magnesium
Aluminium
Carbon
Zinc
Iron
Tin
Lead
Gold   Less reactive

   (i)   Write down the name of a metal that can be extracted from its ore by carbon.

....

(1 mark)

   (ii)   What method is used to extract magnesium or aluminium from their ores?

....

(1 mark)

   (b)   Iron is extracted from iron ore by reduction in the blast furnace.

   (i)   Why is hot air blasted into the furnace?

....

(1 mark)

   (ii)   What is the main impurity in iron ore?

....

(1 mark)

   (iii)   Describe how this impurity is removed in the blast furnace.

....

....

....

(3 marks)

Extracting Aluminium — Electrolysis

The other way of extracting metals from ores is by electrolysis.

A *Molten State* is needed for *Electrolysis*

1) Aluminium is more reactive than carbon so it has to be extracted from its ore by electrolysis.
2) The basic ore is bauxite, and after mining and purifying a white powder is left.
3) This is pure aluminium oxide, Al_2O_3, which has a very high melting point of over 2000°C.
4) For electrolysis to work a molten state is required, but heating to 2000°C would be expensive.

So to lower the costs...

Cryolite is used to *lower the temperature*

1) Instead the aluminium oxide is dissolved in molten cryolite (a less common ore of aluminium).
2) This brings the temperature required down to about 900°C, which makes it much cheaper and easier.
3) The electrodes are made of graphite (carbon).
4) The graphite anode (+ve) does need replacing quite often. It keeps reacting to form CO_2.

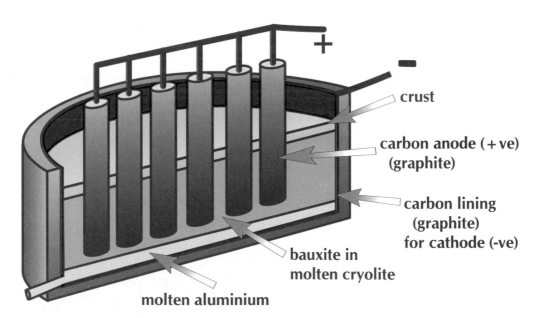

+

–

crust

carbon anode (+ve)
(graphite)

carbon lining
(graphite)
for cathode (-ve)

bauxite in
molten cryolite

molten aluminium

Now you're ready to do some electrolysis.

It's got to be runny

Remember, electrolysis can only take place when the electrolyte (the ore) is either molten or dissolved in something. The ions in the electrolyte need to be able to move, and they can't do this in a solid. Learn about cryolite. You may get asked about how aluminium manufacturers reduce costs.

Extracting Aluminium — Electrolysis

Electrolysis — turning IONS into the ATOMS you want

This is the <u>main object of the exercise</u>:

1) Make the aluminium oxide <u>molten</u> to <u>release</u> the aluminium <u>ions</u>, Al^{3+}, so they're <u>free to move</u>.

2) Stick <u>electrodes</u> in — so that the <u>positive</u> Al^{3+} <u>ions</u> will head straight for the <u>negative electrode</u> (the cathode).

3) At the negative electrode they just can't help picking up some of the <u>spare electrons</u> and "<u>zup</u>", they've turned into aluminium <u>atoms</u> and they <u>sink to the bottom</u>.

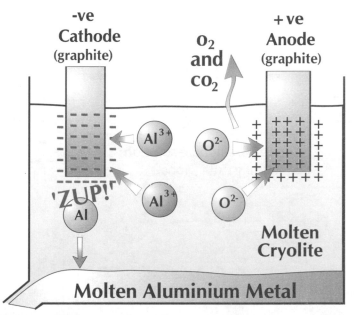

Overall, this is a <u>REDOX (reduction and oxidation) reaction</u>.
You need to know the <u>reactions</u> at both electrodes:

At the Cathode (–ve):	At the Anode (+ve):
$Al^{3+} + 3e^- \rightarrow Al$	$2O^{2-} \rightarrow O_2 + 4e^-$
(<u>Reduction</u> — a gain of electrons)	(<u>oxidation</u> — a loss of electrons)

Electrolysis is Expensive — it's all that electricity...

1) Electrolysis uses <u>a lot of electricity</u> and that can make it pretty <u>expensive</u>.

2) Aluminium smelters usually have <u>their own</u> hydro-electric power station <u>nearby</u> to make the electricity as <u>cheap</u> as possible.

3) Energy is also needed to <u>heat</u> the electrolyte mixture to 900ºC. This is expensive too.

4) The <u>disappearing anodes</u> need frequent <u>replacement</u>. That costs money as well.

5) But in the end, aluminium now comes out as a <u>reasonably cheap</u> and <u>widely-used</u> metal. <u>A hundred years ago</u> it was a very <u>rare</u> metal, simply because it was so <u>hard to extract</u>.

Half equations are a bit tricky — but you need to know them

Don't get fazed by the funny looking half equations with their electrons showing. All they represent is how the ions in the aluminium ore lose or gain electrons at the anode or cathode, so they can become atoms again. Don't forget to balance them.

Purifying Copper by Electrolysis

1) Aluminium is a <u>very reactive metal</u> and <u>has</u> to be removed from its ore by <u>electrolysis</u>.

2) <u>Copper</u> is a very <u>unreactive</u> metal. Not only is it below carbon in the reactivity series, it's also below <u>hydrogen</u>, which means that copper doesn't even react with <u>water</u>.

3) So copper is obtained <u>very easily</u> from its ore by <u>reduction</u> with <u>carbon</u>.

Very pure copper is needed for *electrical* conductors

1) The copper produced by <u>reduction isn't pure enough</u> for use in <u>electrical conductors</u>.

2) The <u>purer</u> it is, the better it <u>conducts</u>. <u>Electrolysis</u> is used to obtain <u>very pure copper</u>.

The **cathode** starts as a <u>thin</u> piece of <u>pure copper</u> and more pure copper <u>adds</u> to it.

The **anode** is just a big lump of <u>impure copper</u>, which will <u>dissolve</u> in the copper(II) sulphate solution.

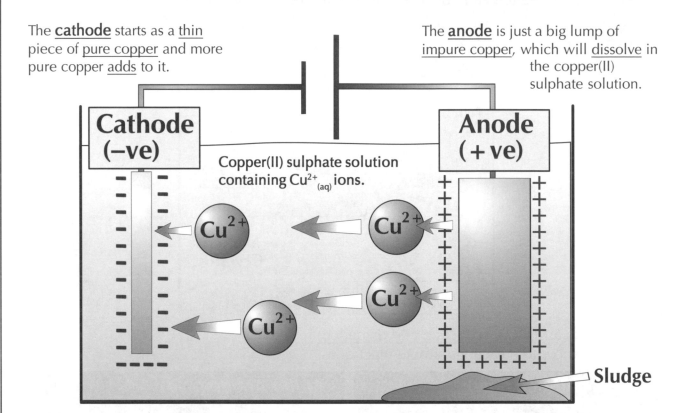

Cathode (–ve)

Copper(II) sulphate solution containing $Cu^{2+}_{(aq)}$ ions.

Cu^{2+} Cu^{2+} Cu^{2+} Cu^{2+}

Anode (+ve)

Sludge

Pure copper is deposited on the pure cathode (–ve)

Copper dissolves from the impure anode (+ve)

The reaction at the <u>cathode</u> is:
$$Cu^{2+}_{(aq)} + 2e^- \rightarrow Cu_{(s)}$$
(reduction)

The reaction at the <u>anode</u> is:
$$Cu_{(s)} \rightarrow Cu^{2+}_{(aq)} + 2e^-$$
(oxidation)

The <u>electrical supply</u> acts by:

1) <u>Pulling electrons off</u> copper atoms at the <u>anode</u>, causing them to go into solution as <u>Cu^{2+} ions</u>.

2) Then <u>offering electrons</u> at the <u>cathode</u> to nearby <u>Cu^{2+} ions</u> to turn them back into <u>copper atoms</u>.

3) The <u>impurities</u> are dropped at the <u>anode</u> as a <u>sludge</u>, whilst <u>pure copper atoms</u> bond to the <u>cathode</u>.

4) The electrolysis can go on for <u>weeks</u> and the cathode is often <u>twenty times bigger</u> at the end of it.

Electrolysis can take an... awfully... long... time...
This is a pretty easy page to learn. Write a mini-essay about purifying copper. Don't forget to learn the diagram and the equations. I know it's not much fun, but life is not all cakes and ale.

Uses of the Three Common Metals

Metals are a lot more interesting than most people ever realise.

*Iron is made into **steel** which is **cheap and strong***

Iron and steel:

Advantages:	Cheap and strong.
Disadvantages:	Heavy, and prone to rusting away.

Iron and steel are used for:

1) Construction of things such as bridges and buildings.
2) Cars and lorries and trains and boats and NOT PLANES.
3) Stainless steel doesn't rust and is used for pans and for fixtures on boats.

Steel may rust and it may not be exactly "space age", but it's strong and it's awfully cheap, and it still has a lot of uses. They make cars out of it for one thing... but gone are the halcyon days when car bodies were hand-crafted from ash frames and lovingly honed to perfection. Now they just shovel them out of big presses by the million.

*Aluminium is **light**, **strong** and **corrosion-resistant***

Strictly speaking you shouldn't say it's "light", you should say it has "low density". Anyway, it's a lot easier to lift and move around than iron or steel.

Useful Properties:

1) Lightweight. (*"low density"*)
2) Can be bent and shaped (for making car body panels, etc.)
3) Strong and very rigid when required.
4) Doesn't corrode due to the protective layer of oxide which always quickly covers it.
5) It's also a good conductor of heat and electricity.

Drawbacks: Not as strong as steel and a bit more expensive.

Common uses:

1) Ladders.
2) Aeroplanes.
3) Range Rover body panels.
4) Drink cans — better than tin-plated steel ones, which can rust if damaged.
5) Greenhouses and window frames.
6) Big power cables used on pylons.

***Copper**: good conductor, easily bent and doesn't corrode*

This is a winning combination which makes it ideal for:

1) Water pipes and gas pipes, because it can be bent to shape by hand without fracturing.
2) Electrical wiring because it can be easily bent round corners and it conducts really well.
3) Forms useful non-corroding alloys such as brass (for trumpets) and bronze (for statues).

Drawbacks: Copper is quite expensive and is not strong.

The Exciting Properties of Metals
What? Some chemistry which is useful in your everyday life? Well, maybe it's only really important if you plan to build your own steam engine or rocket or something, but you'd best learn it anyway.

Warm-Up and Worked Exam Questions

These warm-up questions should ease you gently in and make sure you've got the basics straight. If there's anything you've forgotten, check up on the facts before you do the exam questions.

Warm-up Questions

1) What metal is used to make aeroplane bodies and wings?
2) Which is more reactive, aluminium or copper?
3) What is the name of the most common ore of aluminium?
4) Why is iron used to make bridges?

Worked Exam Question

Wow, an exam question — with the answers helpfully written in. It must be your lucky day.

1   This question is about copper, its purification and its uses.

   (a) Copper is purified by electrolysis.

      (i)   What would you see happening to the pure copper electrode during electrolysis?

      It would get bigger.

      (1 mark)

      (ii) What would you see happening to impurities from the lump of impure copper during electrolysis?

      They would fall to the bottom as sludge

      (1 mark)

      (iii) Finish these ionic equations.

      At the impure electrode: $Cu_{(s)} \rightarrow$ *$Cu^{2+}_{(aq)}$* $+ 2e^-$

      At the pure electrode: $Cu^{2+}_{(aq)} +$ *$2e^-$* $\rightarrow Cu_{(s)}$

      (1 mark)

      (iv) What is the copper sulphate solution for?   *This bit's tricky. Remember, it's all about the flow of ions.*

      It allows Cu^{2+} ions to flow between the electrodes without having to use molten copper which is expensive to heat.

      (2 marks)

   b)   What properties of copper make it suitable for use in water pipes?

      Copper is ductile and malleable. It can be bent into shape easily. It doesn't corrode.

      (2 marks)

Exam Questions

1   Aluminium is separated from aluminium oxide by electrolysis.

a)   Why is a molten state required for the electrolyte?

....

....

(2 marks)

b)   The electrolyte used is aluminium oxide dissolved in molten cryolite, rather than just molten aluminium oxide. Explain why this is.

....

....

....

(2 marks)

c)   What is the melting temperature of the electrolyte?

....

(1 mark)

d)   i)   Write down the ionic equation showing what happens at the cathode.

....

(2 marks)

ii)   Write down the ionic equation showing what happens at the anode.

....

(2 marks)

iii)   Explain why the graphite anode needs to be replaced frequently.

....

....

(2 marks)

Exam Questions

2   Copper used for electrical conductors is extracted from copper ore by reduction with carbon, and is purified by electrolysis.

   (a) Why is the copper produced by reduction not used for electrical conductors before it is purified?

   ....
   (1 mark)

   (b) This diagram shows the purification of copper by electrolysis.

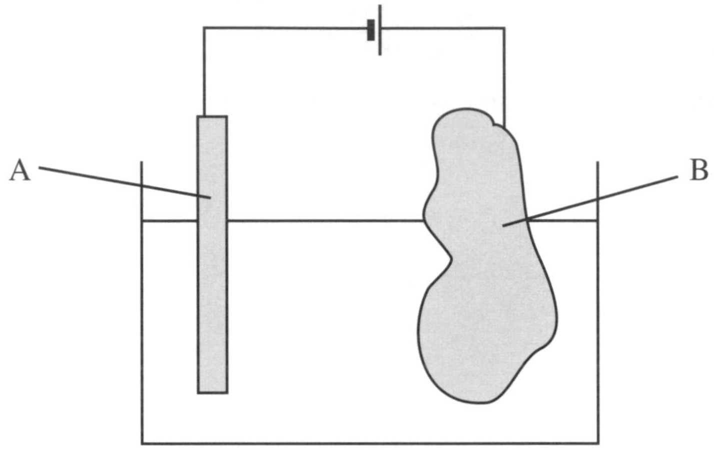

      (i)   Label the cathode and the anode.

            A = ...

            B = ...
            (2 marks)

      (ii) Which electrode gets gradually bigger as the electrolysis goes on?

            ....
            (1 mark)

3   Aluminium is the most common element in the Earth's crust. However, before the late 1900s, aluminium was almost as expensive as silver.
    Using your knowledge of the reactivity series, explain why aluminium was once so expensive and little used.

   ....

   ....

   ....

   ....

   ....
   (3 marks)

Five Uses of Limestone

Limestone is a sedimentary rock, formed mainly from sea shells. It is mostly calcium carbonate.

1) Limestone Used as a Building Material

1) It's great for making into blocks for building with.
   Fine old buildings like cathedrals are often made purely
   from limestone blocks. Acid rain can be a problem though.
2) It's used for statues and fancy carved bits on nice buildings.
   But acid rain is even more of a problem for these.
3) It can also be crushed up into chippings and used for road surfacing.

2) Limestone for Neutralising Acid in lakes and soil

1) Ordinary limestone ground into powder can be used to neutralise acidity in lakes
   caused by acid rain. It can also be used to neutralise acid soils in fields.
2) It works better and faster if it's turned into slaked lime first:

Turning Limestone into Slaked Lime: first heat it up, then add water

1) The limestone, which is mostly calcium carbonate, is heated and it turns into calcium oxide (CaO):

$$\text{limestone} \xrightarrow{\text{HEAT}} \text{quicklime} \quad \text{or} \quad CaCO_3 \xrightarrow{\text{HEAT}} CaO + CO_2$$

2) Calcium oxide reacts violently with water to produce calcium hydroxide (or slaked lime):

$$\text{quicklime} + \text{water} \longrightarrow \text{slaked lime} \quad \text{or} \quad CaO + H_2O \longrightarrow Ca(OH)_2$$

3) Slaked lime is a white powder and can be applied to fields just like powdered limestone.
4) The advantage is that slaked lime acts much faster at reducing the acidity.

3) Limestone and Clay are Heated to Make Cement

1) Clay contains aluminium and silicates and is dug out of the ground.
2) Powdered clay and powdered limestone are roasted in a rotating kiln to produce
   a complex mixture of calcium and aluminium silicates called cement.
3) When cement is mixed with water a slow chemical reaction takes place.
4) This causes the cement to gradually set hard.
5) Cement is usually mixed with sand and chippings to make concrete.
6) Concrete is a very quick and cheap way of constructing buildings — and it shows.

4) Glass is made by melting Limestone, Sand and Soda

1) Just heat up limestone (calcium carbonate) with sand (silicon dioxide) and
   soda (sodium carbonate) until it melts.
2) When the mixture cools it comes out as glass. It's as easy as that.

5) Limestone removes impurities during iron extraction

Limestone is used in blast furnaces. It reacts with impurities to form slag. See pages 40-41.

Limestone's amazingly useful

Limestone's great for lots of things, but there's a down side — ripping open huge quarries spoils the countryside, especially if the limestone's taken from rare 'limestone pavements'.

Making Ammonia: The Haber Process

This is an <u>important industrial process</u>. It produces <u>ammonia</u> (NH_3), which is needed for making <u>fertilisers</u>.

Nitrogen and Hydrogen are needed to make Ammonia

1) The <u>nitrogen</u> is obtained easily from the <u>air</u>, which is <u>78% nitrogen</u> (and 21% oxygen).

2) The <u>hydrogen</u> is obtained from <u>water</u> (steam) and <u>natural gas</u> (methane, CH_4). The methane and steam are <u>reacted together</u> like this:

$$CH_{4\,(g)} + H_2O_{\,(g)} \rightarrow CO_{\,(g)} + 3H_{2\,(g)}$$

3) Hydrogen can also be obtained from <u>crude oil</u>.

The Haber Process is a Reversible Reaction:

$$N_{2\,(g)} + 3H_{2\,(g)} \rightleftharpoons 2NH_{3\,(g)} \quad (+\ heat)$$

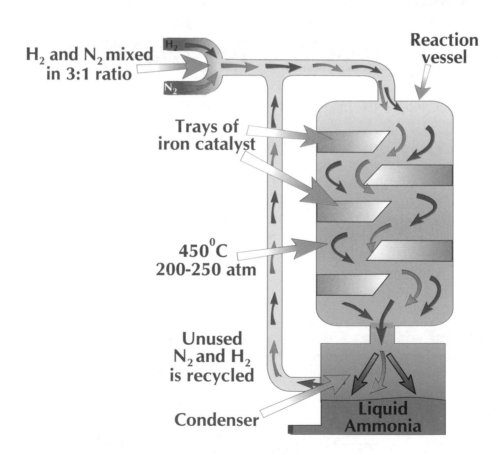

H₂ and N₂ mixed in 3:1 ratio

Reaction vessel

Trays of iron catalyst

450^0C
200-250 atm

Unused N_2 and H_2 is recycled

Condenser

Liquid Ammonia

Pay attention, this is important...

The Haber Process is a biggy, and the Examiners like to ask lots of questions on it. You absolutely MUST memorise the formula for the reversible reaction. It WILL come up in the Exam, and you'll need to know it before you can even think about tackling those big three-mark monsters at the end.

Making Ammonia: The Haber Process

This page is really <u>important</u>. You need to be able to say what will happen if you <u>change the conditions</u> under which the Haber Process is carried out.

Industrial conditions:

<u>Pressure:</u>   **200-250 atmospheres**
<u>Temperature:</u>   **450ºC**
<u>Catalyst:</u>   **Iron**

BECAUSE THE REACTION IS REVERSIBLE, THERE'S A COMPROMISE TO BE MADE:

1) <u>Higher pressures</u> favour the *forward* reaction, hence the <u>200-250 atmospheres</u> operating pressure.

2) However, it turns out that <u>lower</u> temperatures improve the forward reaction. At least it does in terms of the <u>proportion</u> of hydrogen and nitrogen converting to ammonia. This is called the <u>yield</u>.

3) The trouble is, <u>lower temperatures</u> mean a <u>slower rate of reaction</u>. (This is different from <u>yield</u>. The same yield will be achieved but it will take longer.)

4) So the 450ºC is a <u>compromise</u> between <u>maximum yield</u> and <u>speed of reaction</u>.

5) The pressure used is also a compromise. Even <u>higher</u> pressures would <u>increase</u> the yield further, but the plant would be <u>more expensive to build</u>. In the end it all comes down to <u>minimising costs</u>.

Extra Notes:

1) The hydrogen and nitrogen are mixed together in a <u>3:1 ratio</u>.

2) Because the reaction is <u>reversible</u>, not all of the nitrogen and hydrogen will <u>convert</u> to ammonia.

3) The <u>ammonia</u> is formed as a <u>gas</u> but as it cools in the condenser it <u>liquefies</u> and is <u>removed</u>.

4) The N_2 and H_2 which didn't react are <u>recycled</u> and passed through again so <u>none is wasted</u>.

There's more about the Haber Process on pages 171-173

200 atmospheres? — that could give you a headache

There are quite a lot of details on these two pages. They're pretty keen on the Haber process in the Exams so you'd be well advised to learn all this. They could easily test you on any of these details. Use the same good old method: <u>Learn it, cover it up, repeat it back to yourself, check, try again...</u>

Using Ammonia to Make Fertilisers

1) Ammonia can be *Oxidised* to form *Nitric Acid*

There are two stages to this reaction:

a) Ammonia gas reacts with *oxygen* over a *hot platinum catalyst*:

$$4NH_{3\,(g)} + 5O_{2\,(g)} \rightarrow 4NO_{(g)} + 6H_2O_{(g)}$$

This first stage is very exothermic and produces its own heat to keep it going.
The nitrogen monoxide must be cooled before the next stage, which happens easily:

b) The *nitrogen monoxide* reacts with *water and oxygen*...

$$6NO_{(g)} + 3O_{2\,(g)} + 2H_2O_{(g)} \rightarrow 4HNO_{3\,(g)} + 2NO_{(g)}$$

...to form nitric acid, HNO_3

The nitric acid produced is very useful for other chemical processes.
One such use is to make ammonium nitrate fertiliser...

2) Ammonia can be neutralised with *Nitric Acid* ...

to make *Ammonium Nitrate* fertiliser

This is a straightforward neutralisation reaction between an alkali (ammonia) and an acid.
The result is of course a neutral salt:

$$NH_{3\,(g)} + HNO_{3\,(aq)} \rightarrow NH_4NO_{3\,(aq)}$$
Ammonia + Nitric acid → Ammonium nitrate

Ammonium nitrate is an especially good fertiliser because it has nitrogen from two sources,
the ammonia and the nitric acid. Kind of a double dose. Plants need nitrogen to make proteins.

Excessive Nitrate Fertiliser causes *Eutrophication* and *Health Problems*

1) If nitrate fertilisers wash into streams they set off a cycle of mega-growth, mega-death and mega-decay. Plants and green algae grow out of control, then start to die off because there's too many of them. Then bacteria take over, feeding off the dying plants and using up the oxygen in the water. Then the fish die because they can't get enough oxygen. It's called eutrophication (See the Biology Book for more details).

2) If too many nitrates get into drinking water it can cause health problems, especially for young babies. Nitrates prevent the blood from carrying oxygen properly and children can turn blue and even die.

3) To avoid these problems it's important that artificial nitrate fertilisers are applied carefully by all farmers — they must take care not to apply too much, and not if it's going to rain soon.

You've just got to learn it I'm afraid
Basically, this page is about how ammonia is turned into ammonium nitrate fertiliser.
More equations to learn I'm afraid, but in Chemistry it was ever thus.

Warm-Up and Worked Exam Questions

Questions about ammonia and fertilisers will either ask you to write down the equations for making ammonia and ammonium nitrate, or they'll ask you to describe the Haber Process.

Warm-up Questions

1) What is limestone mixed with to make cement?
2) Why is slaked lime used on fields and lakes?
3) Is ammonia a base or an acid?
4) What is used as a catalyst in the Haber Process?
5) Name the environmental problem caused by nitrate fertilisers washing into streams.

Worked Exam Question

This worked exam question is about the Haber Process. Have a look through it, and make sure you would know what to write about.

1   This question is about how ammonia is made from nitrogen and hydrogen.

(a) What is the source of the nitrogen used in the Haber Process?

The air

(1 mark)

(b) What is the source of the hydrogen used in the Haber Process?

A reaction between steam and natural gas

(1 mark)

(c) (i) Fill in this equation to show the reaction between nitrogen and hydrogen.

$$N_2 + 3H_2 \rightleftharpoons 2NH_3$$

Make sure it's a balanced equation.

(2 marks)

(ii) What ratio of hydrogen to nitrogen is pumped into the reactor?

3:1

That's easy if you've got the balanced equation right — it's the numbers in front of the H_2 and N_2

(1 mark)

(d) The reaction between nitrogen and hydrogen is reversible.

(i) What happens to unreacted hydrogen and nitrogen?

It is passed through the reactor again

(1 mark)

(ii) What conditions favour a high **yield** of ammonia?

High pressure

Low temperature

There are two marks for this, so you need two conditions. Remember it's low temperature.

(2 marks)

Exam Questions

1   This question is about the uses of limestone.

   a)   Name the two compounds which are mixed with limestone and heated to make glass.

   ....

   (1 mark)

   b)   Limestone is mostly calcium carbonate.

   Complete this equation to show the reaction which takes place when limestone is heated.

   $CaCO_3 \rightarrow$ .... + ...

   (2 marks)

2   Ammonia is made from nitrogen and hydrogen using the Haber Process.

   a)   i)   Describe the industrial conditions used in the Haber Process

   ....

   ....

   (3 marks)

      ii)   Lower temperatures produce higher yields of ammonia. Why is the temperature used in the Haber Process so high?

   ....

   ....

   (2 marks)

   b)   Ammonia is reacted with acid to make fertiliser.

      i)   Write a balanced equation for the reaction between nitric acid and ammonia.

   ....

   (2 marks)

      ii)   Describe what happens when excess nitrate fertiliser washes into streams and rivers.

   ....

   ....

   ....

   (3 marks)

Revision Summary

Section Two is pretty interesting stuff. Relatively speaking. Anyway, whether it is or it isn't, the only thing that really matters is whether you've learnt it all or not. These questions aren't exactly friendly, but they're a serious way of finding out what you don't know. And don't forget, that's what revision is all about — finding out what you don't know and then learning it till you do. Practise these questions as often as necessary. Your ultimate aim is to be able to answer all of them easily.

1) Describe how crude oil is formed. What length of time did it take?
2) What does crude oil consist of?
3) Draw the full diagram showing the fractional distillation of crude oil.
4) What are the seven main fractions obtained from crude oil, and what are they used for?
5) What are hydrocarbons? Describe four properties and how they vary with the molecule size.
6) Give the equations for complete and incomplete combustion of hydrocarbons.
7) Which type is dangerous and why? What are the flames for these two types of combustion?
8) What is "cracking"? Why is it done?
9) Give a typical example of a substance which is cracked and the products that you get.
10) What are the industrial conditions used for cracking?
11) What are alkanes and alkenes? What is the basic difference between them?
12) Draw the structures of the first four alkanes and the first three alkenes and give their names.
13) List four differences in the chemical properties of alkanes and alkenes.
14) What are polymers? What kind of substances can form polymers?
15) Draw diagrams to show how ethene, propene and styrene form polymers.
16) Name four types of plastic, give their physical properties and say what they're used for.
17) What are rocks, ores and minerals? Name one metal found as a metal rather than an ore?
18) What are the two methods for extracting metals from their ores?
19) What decides which method is needed?
20) Draw a diagram of a blast furnace. What are the three raw materials used in it?
21) Write down the equations for how iron is obtained from its ore in the blast furnace.
22) What is slag? Write two equations for the formation of slag, and give two uses of it.
23) How is aluminium extracted from its ore? Give four operational details and draw a diagram.
24) Give the two half equations for this process.
25) Explain three reasons why this process is so expensive.
26) How is copper extracted from its ore? How is it then purified, and why does it need to be?
27) Draw a diagram for the purifying process and give the two equations.
28) Describe the advantages and disadvantages of iron (and steel), and give six uses for it.
29) Describe the advantages and disadvantages of aluminium, and give six uses for it.
30) Describe the advantages and disadvantages of copper, and give three uses for it.
31) What are the five main uses of limestone?
32) Give the equations for turning limestone into slaked lime. Why is this useful?
33) Give four details about what cement is made of and how it works.
34) What is the Haber process? What are the raw materials for it and how are they obtained?
35) Draw a full diagram for the Haber process and explain the temperature and pressure used.
36) Give full details of how ammonia is turned into nitric acid, including equations.
37) What is the main use of ammonia? Give the equation for producing ammonium nitrate.
38) Give two problems resulting from nitrate fertilisers. Explain fully what "eutrophication" is.

Nine Types of Chemical Change

There are <u>nine</u> types of chemical change you should know about. It's worth learning them <u>here and now</u>.

1) THERMAL DECOMPOSITION — *breakdown on heating*

This is when a substance <u>breaks down</u> into simpler substances <u>when heated</u>, often with the help of a <u>catalyst</u>. It's different from a reaction because there's only <u>one substance</u> to start with. <u>Cracking of hydrocarbons</u> is a good example of thermal decomposition.

2) NEUTRALISATION — *acid + alkali gives salt + water*

This is simply when an <u>acid</u> reacts with an <u>alkali</u> (or base) to form a <u>neutral</u> product, which is neither acid nor alkali (usually a <u>salt</u> solution).

3) DISPLACEMENT — *one element kicking another one out*

This is a reaction where a <u>more reactive</u> element reacts with a compound and <u>pushes out</u> a <u>less reactive</u> "rival" element. <u>Metals</u> are the most common example. Magnesium will react with iron sulphate to push the iron out and form magnesium sulphate.

4) PRECIPITATION — *solid forms in solution*

This is a reaction where <u>two solutions react</u> and a <u>solid</u> forms in the solution and <u>sinks</u>. The solid is said to "<u>precipitate out</u>" and, confusingly, the solid is also called "<u>a precipitate</u>".

5) OXIDATION — *loss of electrons*

<u>Oxidation</u> is the <u>addition of oxygen</u>. Iron becoming iron oxide is oxidation. The more technical and general definition of oxidation is "<u>the loss of electrons</u>".

> Remember
> *"OIL RIG"*
> *(Oxidation Is Loss,*
> *Reduction Is Gain)*
> *of electrons, NOT oxygen*

6) REDUCTION — *gain of electrons*

<u>Reduction</u> is the <u>reverse of oxidation</u>, i.e. the <u>loss of oxygen</u>. Iron oxide is <u>reduced</u> to iron. The more technical and general definition of reduction is "<u>the gain of electrons</u>". Note that <u>reduction</u> is <u>gain</u> of electrons. That's the way to remember it — it's kind of <u>the wrong way round</u>.

7) EXOTHERMIC REACTIONS — *give out heat*

Exothermic reactions <u>give out energy</u>, usually as heat. "Exo-" as in "Exit", or "out". Any time a <u>fuel burns</u> and <u>gives off heat</u> it's an <u>exothermic</u> reaction.

8) ENDOTHERMIC REACTIONS — *take in heat*

<u>Endothermic</u> reactions need heat <u>putting in</u> constantly to make them work. Heat is needed to <u>break chemical bonds</u>. The <u>products</u> of endothermic reactions are likely to be <u>more useful</u> than the <u>reactants</u>, otherwise we <u>wouldn't bother putting all the energy in</u>, e.g. turning <u>iron oxide</u> into <u>iron</u> is an endothermic process. We need a lot of heat from the coke to keep it happening.

9) REVERSIBLE REACTIONS — *they go both ways*

<u>Reversible</u> reactions are ones that will cheerfully go in <u>both</u> directions at the <u>same time</u>. In other words, the <u>products</u> can easily turn back into the <u>original reactants</u>.

Come on, there are only nine of them

<u>A nice easy page to learn</u>. You should know a lot of this already. All the same, it wouldn't hurt you to cover the page and write out a brief definition of each chemical change from memory.

Balancing Equations

Equations need a lot of practice if you're going to get them right.

This is just a reminder of the basics.

But every time you do an equation you need to practise getting it right rather than skating over it.

The **Symbol Equation** shows the atoms on both sides:

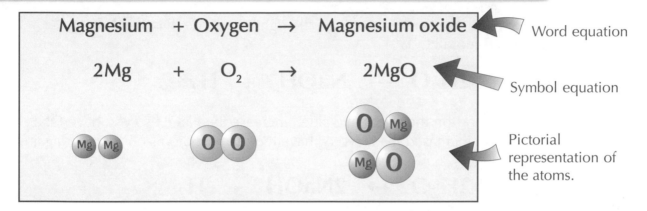

Magnesium	+ Oxygen	→	Magnesium oxide
2Mg	+ O$_2$	→	2MgO

Word equation

Symbol equation

Pictorial representation of the atoms.

Balancing The Equation — match them up **one by one**

1) There must always be the <u>same</u> number of atoms on <u>both sides</u> — they can't just <u>disappear</u>.

2) You <u>balance</u> the equation by putting numbers <u>in front</u> of the formulae where needed. Take this equation for reacting sodium with water:

$$Na + H_2O \rightarrow NaOH + H_2$$

The <u>formulae</u> are all correct but the numbers of some atoms <u>don't match up</u> on both sides. You <u>can't change formulae</u> like H_2O to H_2O_2. You can only put numbers <u>in front of them</u>.

Method: Balance just **ONE type of atom** at a time

The more you practise, the quicker you get, but all you do is this:

1) Find an element that <u>doesn't balance</u> and <u>pencil in a number</u> to try and sort it out.

2) <u>See where it gets you</u>. It may create <u>another imbalance</u> but pencil in <u>another number</u> and see where that gets you.

3) Carry on chasing <u>unbalanced</u> elements and it'll <u>sort itself out</u> pretty quickly.

Take a look at the examples over the page

Being able to balance is really useful
Balancing equations is really easy to learn, and worth an awful lot of marks in the Exam.
Learn the method above, and go over the page to see how it's applied.

Balancing Equations

Here's the formula from the previous page for the reaction between sodium and water:

$$Na + H_2O \rightarrow NaOH + H_2$$

1) You'll notice that there are more H atoms on the right hand side than there are on the left. You'll need more H_2O:

$$Na + 2H_2O \rightarrow NaOH + H_2$$

2) Now you're one H short on the right hand side. Increase the NaOH so you have ONE more H atom. (If you increased the H_2 you'd have too many H atoms on the right again):

$$Na + 2H_2O \rightarrow 2NaOH + H_2$$

3) But now you need more Na on the left hand side:

$$2Na + 2H_2O \rightarrow 2NaOH + H_2$$

4) And suddenly there it is. Everything balances. It took a while, but by being methodical, we got there in the end.

State Symbols tell you what Physical State it's in

These are easy enough, just make sure you know them, especially aq (aqueous).

(s) — Solid
(l) — Liquid
(g) — Gas
(aq) — Dissolved in water

E.g. $2Mg_{(s)} + O_{2(g)} \rightarrow 2MgO_{(s)}$

It's not that hard once you get going

Just read over the example again, and have a go at balancing the equation below before you move on. Add state symbols too: Iron(III) oxide + hydrogen → iron + water

Electrolysis and the Half Equations

Electrolysis means "Splitting Up with Electricity"

1) It requires a liquid, called the <u>electrolyte</u> which will <u>conduct electricity</u>.

2) Electrolytes are usually <u>free ions dissolved in water</u>, e.g. <u>dilute acids</u> like HCl, and <u>dissolved salts</u>, e.g. NaCl solution.

3) Electrolytes can also be <u>molten ionic substances</u>, but this involves <u>higher temperatures</u>. In either case it's the <u>free ions</u> which <u>conduct</u> the electricity and allow the whole thing to work.

4) The electrical supply acts like an <u>electron pump</u>, taking electrons <u>away from</u> the +ve anode and pushing them <u>onto the –ve cathode</u>. Ions <u>gain or lose</u> electrons at the electrodes and <u>neutral atoms and molecules</u> are released.

NaCl dissolved

Molten NaCl

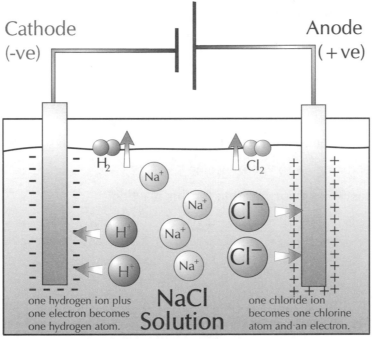

Cathode (-ve) Anode (+ve)

<u>Metals</u> will always be produced at the <u>cathode</u> because metals form <u>+ve ions</u>.

+ve ions are called CATIONS because they're attracted to the –ve cathode.

<u>Hydrogen</u> is also produced at the <u>–ve cathode</u>.

one hydrogen ion plus one electron becomes one hydrogen atom.

NaCl Solution

one chloride ion becomes one chlorine atom and an electron.

<u>ALL Non-metals</u> (except hydrogen) have <u>–ve ions</u> and so they'll be produced at the <u>+ve anode</u>.

–ve ions are called ANIONS because they're attracted to the <u>anode</u>.

In this solution, hydrogen gas forms (from H^+ ions in the water) rather than sodium metal, because sodium is too reactive to form.

The **Half Equations** — *make sure* **the electrons balance**

The main thing is to make sure the <u>number of electrons</u> is the <u>same</u> for <u>both half equations</u>. For the above cell the basic half equations are:

<u>Cathode:</u> $H^+_{(aq)} + e^- \rightarrow H$

<u>Anode:</u> $Cl^-_{(aq)} \rightarrow Cl + e^-$

These equations <u>aren't finished</u> because both the hydrogen and the chlorine come off as <u>gases</u>. They must be <u>rewritten</u> with H_2 and Cl_2, like this:

<u>Cathode:</u> $2H^+_{(aq)} + 2e^- \rightarrow H_{2(g)}$

<u>Anode:</u> $2Cl^-_{(aq)} \rightarrow Cl_{2(g)} + 2e^-$

Note that there are <u>two electrons</u> in <u>both</u> half equations, which means they're nice and <u>balanced</u>. This gives the <u>overall equation</u>:

$$2HCl_{(aq)} \rightarrow H_{2(g)} + Cl_{2(g)}$$

Half Equations are fiddly, but they need learning

Electrolysis can be a bit confusing. Make an effort to learn all the details, especially how the two half equations are really just <u>one</u> equation, but it kind of happens in two places, <u>joined by a battery</u>.

Warm-Up and Worked Exam Questions

Balancing equations is incredibly important. The best way to learn how to do it is to practise.

Warm-up Questions

1) What's the term for reactions that take in heat from their surroundings?
2) What kind of reaction is this?

$$HNO_{3(aq)} + NaOH_{(aq)} \rightarrow NaNO_{3(aq)} + H_2O$$

3) What's the term for reactions where two solutions react to produce a solid which sinks?
4) Are anions positive or negative?
5) Sodium chloride solution can be electrolysed. Write down the ionic half equation showing what happens at the cathode.

Worked Exam Question

You can be asked to write a balanced equation in questions about all kinds of topics, so pay extra special attention when you read through this worked example.

1   Equations are used to represent chemical reactions.

(a) This is a word equation for the decomposition of hydrogen peroxide (H_2O_2).

hydrogen peroxide $\rightarrow$ oxygen + water.

Write a balanced symbol equation for this reaction.

4 hydrogens and 4 oxygens on this side... $2H_2O_2 \rightarrow O_2 + 2H_2O$ *... and 4 hydrogens and 4 oxygens on this side*

(2 marks)

(b) Potassium hydroxide reacts with sulphuric acid to produce a salt and water. Write a balanced symbol equation to show this.

$$2KOH_{(aq)} + H_2SO_{4(aq)} \rightarrow K_2SO_{4(aq)} + 2H_2O$$

(2 marks)

You need 2 potassiums here because it's K_2SO_4 on the other side.

2 potassiums, 6 oxygens, 4 hydrogens, 1 sulphur on both sides. That's balanced.

(c) Ammonia gas reacts with oxygen over a hot platinum catalyst to produce nitrogen monoxide and water. Write a balanced equation to show this.

$$4NH_3 + 5O_2 \rightarrow 4NO + 6H_2O$$

Pencil in a number in front of the NH_3 and try to make it work...

(3 marks)

..2 won't work. You'd need 2NO on the RHS (to balance N) and $3H_2O$ (to balance H), which means you'd need 5 O on the LHS, and that won't go...

...3 won't work. You'd need 3NO on the RHS and 9H — which won't go into H_2O...

...4 works! You need 4NO and $6H_2O$ on the RHS. That means 10 O on the LHS, which means you need $5O_2$. Phew.

Exam Questions

1   This diagram shows sodium bromide solution being electrolysed.

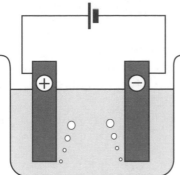

   a)   Can solid sodium bromide be electrolysed?

   ....

   (1 mark)

   b)   At the negative electrode, bubbles of a gas are formed. This gas is collected in a test
         tube, and is found to burn with a squeaky pop. What is this gas, and why is it formed
         at the cathode instead of sodium?

   ....

   ....

   ....

   (2 marks)

   c)   Write down a balanced ionic half equation to show what happens at the positive
         electrode.

   ....

   (1 mark)

2   A solution of copper sulphate is electrolysed using graphite electrodes.

   a)   At the positive electrode, bubbles of oxygen are formed. Describe a test for oxygen.

   ....

   ....

   (1 mark)

   b)   Balance the following ionic half equations to show what happens at the electrodes.

         At the positive electrode: ....$H_2O_{(l)} \rightarrow O_{2(g)} +$ ....$H^+_{(aq)} + 4e^-$

         At the negative electrode: ....$Cu^{2+}_{(aq)} + 4e^- \rightarrow$ ....$Cu_{(s)}$

   (2 marks)

Relative Formula Mass

The biggest trouble with <u>relative atomic mass</u> and <u>relative formula mass</u> is that they <u>sound</u> intimidating. In fact, they're not that complicated at all. Just read this carefully...

Relative Atomic Mass, A_r

1) This is just a way of saying how <u>heavy</u> different atoms are <u>compared to each other</u>.
2) The <u>relative atomic mass</u> A_r is nothing more than the <u>mass number</u> of the element.
3) In the periodic table, the elements all have <u>two</u> numbers. The smaller number is the atomic number (how many protons it has).
But the <u>bigger number</u> is the <u>mass number</u> (how many protons and neutrons it has) which, kind of obviously, is also the <u>relative atomic mass</u>.

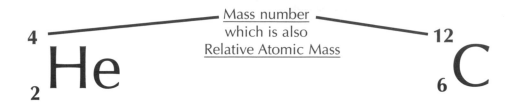

Mass number
which is also
<u>Relative Atomic Mass</u>

$$_2^4He \qquad\qquad _6^{12}C$$

Helium has $A_r = 4$. Carbon has $A_r = 12$. (So carbon atoms are <u>3 times heavier</u> than helium atoms.)

Relative Formula Mass, M_r

If you have a compound like $MgCl_2$ then it has a <u>relative formula mass</u>, M_r, which is just all the relative atomic masses <u>added together</u>.
For $MgCl_2$ it would be:

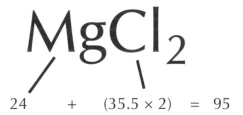

$$MgCl_2$$
$$24 \quad + \quad (35.5 \times 2) \quad = \quad 95$$

So the M_r for $MgCl_2$ is simply <u>95</u>

You can easily get the A_r for any element from the <u>Periodic Table</u>, but in a lot of questions they give you them anyway. Here's another example for you:

QUESTION: *Find the relative formula mass for calcium carbonate, CaCO₃ using the given data:* A_r for Ca = 40 A_r for C = 12 A_r for O = 16

$$CaCO_3$$
$$40 \quad + \quad 12 \quad + \quad (16 \times 3) = 100$$

So the relative formula mass for CaCO₃ is <u>100</u>

And that's all there is to it.

Easy as 2 + 2....

Relative atomic mass is just a case of looking up the right number. Remember that it's always the **LARGER** number. As for relative formula masses, all you need to be able to do is add up right.
Just remember to multiply out any numbers before you add the elements together.

Relative Formula Mass

Although relative atomic mass and relative formula mass are <u>easy enough</u>, it can get just a little bit <u>trickier</u> when you start getting into other calculations which use them. It depends on how good your maths is basically, because it's all to do with ratios and percentages.

Calculating % Mass *of an Element in a Compound*

This is really easy — so long as you've learnt this formula:

$$\text{PERCENTAGE MASS OF AN ELEMENT IN A COMPOUND} = \frac{A_r \times \text{No. of atoms (of that element)}}{M_r \text{ (of whole compound)}} \times 100$$

If you don't learn the formula then you'd better be pretty smart — or you'll struggle.

EXAMPLE

<u>QUESTION:</u> *Find the percentage mass of sodium in sodium carbonate, Na_2CO_3.*

<u>ANSWER:</u>

A_r of sodium = 23
A_r of carbon = 12
A_r of oxygen = 16

So... M_r of Na_2CO_3 = $(2\times23)+12+(3\times16)=106$

Now use the formula:

$$\underline{\text{Percentage mass}} = \frac{A_r \times n}{M_r} \times 100 = \frac{23 \times 2}{106} \times 100 = \mathbf{43.4\%}$$

And there you have it. Sodium represents <u>43.4%</u> of the mass of sodium carbonate.

Playing the percentage game

Remember, working out a percentage mass is just like working out any other percentage. There's no great mystery to it at all, just work out what the whole molecule weighs, and find the relative atomic mass of the element you need. Stick it into the formula and away you go.

The Empirical Formula

Finding The Empirical Formula *(from Masses or Percentages)*

This also sounds a lot worse than it really is. Try this for an easy stepwise method:

1) <u>List all the elements</u> in the compound (there's usually only two or three).
2) <u>Underneath them</u>, write their <u>experimental masses or percentages</u>.
3) <u>Divide</u> each mass or percentage <u>by the A_r</u> for that particular element.
4) Turn the numbers you get into <u>a nice simple ratio</u>
   by multiplying and/or dividing them by well-chosen numbers.
5) Get the ratio in its <u>simplest form</u>, and that tells you the formula of the compound.

EXAMPLE

QUESTION: *Find the empirical formula of the iron oxide produced when 44.8g of iron react with 19.2g of oxygen. (A_r for iron = 56, A_r for oxygen =16)*

ANSWER:

1) List the two elements: Fe   O

2) Write in the experimental masses: 44.8   19.2

3) Divide by the A_r for each element: $\dfrac{44.8}{56} = 0.8$ $\dfrac{19.2}{16} = 1.2$

4) Multiply by 10...   8   12
        ...then divide by 4:   2   3

5) So the simplest formula is 2 atoms of Fe to 3 atoms of O, i.e. **Fe_2O_3**. And that's it done.

> *You need to realise (for the Exam) that this <u>EMPIRICAL METHOD</u> (i.e. based on <u>experiment</u>) is the <u>only way</u> of finding out the formula of a compound. Rust is iron oxide, sure, but is it FeO, or Fe_2O_3? Only an experiment to determine the empirical formula will tell you for certain.*

It's not as hard as it looks...

Finding empirical formulae will only fox you if you let it. By far the easiest thing for you to do is to learn the step-by-step process outlined above, and then apply it to <u>every</u> question about this topic. This is the best foolproof method to make sure you get all the marks you deserve.

Warm-Up and Worked Exam Questions

Without a good warm-up you're likely to strain a brain cell or two. So take the time to run through these simple questions first. Use the periodic table at the front of the book.

Warm-up Questions

Remember, use the Periodic Table at the front of the book to find atomic numbers and mass numbers.

1) How many times heavier is one atom of oxygen than one atom of helium?

2) Write down the number of neutrons in beryllium, sodium and magnesium.

3) Write down the relative formula mass of Na_2CO_3, $(NH_4)_2SO_4$, $Ca(OH)_2$.

4) Find the percentage mass of nitrogen in NH_3, NH_4Cl, NH_4NO_3.

Worked Exam Question

You know the routine by now — work carefully through this example and make sure you understand it. Then it's onto the real test — doing some exam questions for yourself.

1   Glucose ($C_6H_{12}O_6$) is a blood sugar, which breaks down in human tissues to release energy. This process, called aerobic respiration, requires oxygen.

The balanced chemical equation for aerobic respiration is:

$$C_6H_{12}O_6 + 6O_2 \rightarrow 6CO_2 + 6H_2O + Energy$$

a)   Calculate the percentage mass of oxygen in:

180 = the relative formula mass of $C_6H_{12}O_6$

i)   glucose

$(6 \times 12) + (12 \times 1) + (6 \times 16) = 180$ $\dfrac{(6 \times 16)}{180} \times 100 = 53.3\%$

(1 mark)

ii)   carbon dioxide

$12 + (2 \times 16) = 44$ $\dfrac{(2 \times 16)}{44} \times 100 = 72.7\%$

(1 mark)

b)   One molecule of glucose breaks down in the presence of oxygen to produce six molecules of carbon dioxide. Calculate the relative formula mass of the six molecules of carbon dioxide.

$12 + (2 \times 16) = 44$ $6 \times 44 = 264$

(1 mark)

c)   What mass of oxygen is needed in order to completely break down 180g of glucose?

M_r of glucose = 180. M_r of O_2 = $2 \times 16 = 32$

$1 \times 180 = 180g$ $6 \times 32 = 192g$

The chemical equation shows 6 molecules of O_2 are needed for every 1 molecule of glucose...

...so 192g of O_2 are needed for every 180g of glucose.   *(2 marks)*

Exam Questions

1   Ribose is another sugar used by living cells to produce energy. Ribose contains 40.0% carbon, 6.67% hydrogen and 53.33% oxygen by mass.

   (a)   How many grams of carbon, hydrogen and oxygen are in 100g of ribose?

   ....

   (1 mark)

   (b)   Use this information to calculate the empirical formula of ribose.

   ....

   ....

   ....

   (2 marks)

2   The empirical formula for copper(II) sulphate is $CuSO_4$.

   (a)   Calculate the percentage composition of each element in copper (II) sulphate.

   ....

   ....

   ....

   (3 marks)

   (b)   Calculate the mass of copper in 180g of copper(II) sulphate.

   ....

   ....

   ....

   (2 marks)

3   Aspirin has the chemical formula $C_9H_8O_4$. Calculate the percentage mass of carbon in aspirin.

   ....

   ....

   ....

   (2 marks)

Calculating Masses in Reactions

These can be kind of scary, but you'll soon get the hang of it.

The Three Important Steps — not to be missed...

(Miss one out and it'll all go horribly wrong, believe me.)

1) <u>Write out</u> the balanced <u>Equation</u>
2) <u>Work out M</u>$_r$ — just for the <u>two bits you want</u>
3) Apply the rule: <u>Divide to get one, then multiply to get all</u>
   (But you have to apply this first to the substance they give
   information about, and *then* the other one!)

EXAMPLE

What mass of magnesium oxide is produced when 60g of magnesium is burned in air?

1) <u>Write out the balanced equation</u>:

$$2Mg + O_2 \rightarrow 2MgO$$

2) <u>Work out the relative formula masses</u>:

(don't do the oxygen — we don't need it)

$$2 \times 24 \rightarrow 2 \times (24+16)$$
$$48 \rightarrow 80$$

3) Apply the rule: <u>Divide to get one, then multiply to get all</u>

The two numbers, 48 and 80, tell us that *48g of Mg react to give 80g of MgO.*
Here's the tricky bit. You've now got to be able to write this down:

48g of Mg ......reacts to give.....80g of MgO

1g of Mg ......reacts to give.....

60g of Mg .....reacts to give......

<u>The big clue</u> is that in the question they've said we want to burn "<u>60g of magnesium</u>"
i.e. they've told us how much <u>magnesium</u> to have, and that's how you know to write
down the <u>left hand side</u> of it first, because:

We'll first need to ÷ by 48 to get 1g of Mg
and then need to × by 60 to get 60g of Mg.

<u>Then</u> you can work out the numbers on the other side (shown in orange below) by
realising that you must <u>divide both sides by 48</u> and then <u>multiply both sides by 60</u>.
It's tricky.

÷48   48g of Mg .... 80g of MgO   ÷48

×60   1g of Mg .... 1.67g of MgO   ×60

     60g of Mg .... 100g of MgO

You should realise that <u>in
practice</u> 100% yield may not
be obtained in some
reactions, so the amount of
product might be <u>slightly less
than calculated</u>.

This finally tells us that <u>60g of magnesium will produce 100g of magnesium oxide</u>.

If the question had said "Find how much magnesium gives 500g of magnesium oxide," you'd
fill in the MgO side first instead, <u>because that's the one you'd have the information about</u>.

Reaction Mass Calculations need learning now

<u>Learn the three rules</u> at the top of the page and practise the example till you can do it fluently.
Find the mass of calcium which gives 30g of calcium oxide (CaO), when burnt in air.

Calculating Volumes

These are OK as long as you <u>LEARN</u> the formula in the <u>dark blue box</u> and know how to use it.

1) *Calculating the Volume when you know the Masses*

For this type of question there are <u>two stages</u>:

1) <u>Find the reacting mass</u>, exactly like in the example on the last page.

2) Then <u>convert the mass into a volume</u> using this formula:

$$\frac{\text{VOL. OF GAS (in cm}^3)}{24,000} = \frac{\text{MASS OF GAS}}{M_r \text{ of gas}}$$

This formula comes from the well known fact that:

<u>A mass of M$_r$ in grams</u>, of any gas, will always occupy <u>24 litres</u>
(at room temperature and pressure) — and it's the same for <u>any gas</u>.

I reckon it's easier to learn and use the formula, but it's certainly worth knowing that fact too.

Example

Find the volume of carbon dioxide produced (at room temperature and pressure) when 2.7g of carbon is completely burned in oxygen. (A$_r$ of carbon = 12, A$_r$ of oxygen = 16)

1) Balanced equation: $C + O_2 \rightarrow CO_2$

2) Fill in M$_r$ for each: ÷12 12 32 44 ÷12

3) Divide for one, times for all: ×2.7 1 .... 3.6666667 ×2.7
2.7 .... 9.8999999
= 9.9

4) So 2.7g of C gives 9.9g of CO$_2$. Now the new bit:

5) <u>Using the above formula:</u> $\frac{\text{Volume}}{24,000} = \frac{\text{MASS}}{M_r}$ ⇒ $\text{Volume} = \frac{\text{MASS} \times 24,000}{M_r}$

so Volume = (MASS/M$_r$) × 24,000 = (9.9/44) × 24,000 = 5400.

= 5400cm^3 or 5.40 litres

Another really important formula to learn

The most important thing from this page is <u>THE FORMULA</u>. It's not only vital for the stuff on this page, but for the stuff on the next page too.

Calculating Volumes

Once you've mastered the calculations on the last page, this stuff should be second nature. It's almost exactly the same, but in reverse.

2) *Calculating the Mass when you're given the Volume*

For this type of question the two stages are in the reverse order:

1) First find the mass from the volume using the same formula as before:

$$\frac{\text{Vol. of gas (in cm}^3)}{24{,}000} = \frac{\text{MASS OF GAS}}{M_r \text{ of gas}}$$

2) Then find the reacting mass, exactly like in the example on the last page.

EXAMPLE

Find the mass of 6.2 litres of oxygen gas.   (A_r of oxygen = 16)

Using the above formula:

$$\frac{6{,}200}{24{,}000} = \frac{\text{Mass of Gas}}{32}$$

(Look out, 32, because it's O_2 — not O.)

Hence,   Mass of Gas = (6,200÷24,000) × 32 =   [8.2666667]   = 8.27g

The question would likely go on to ask what mass of CO_2 would be produced if this much oxygen reacted with carbon. In that case you would now just apply the same old method from page 69:

So, write out the balanced equation:

$$O_2 \ + \ C \ \rightarrow \ CO_2$$

Find the M_rs: 32   12+32 = 44

So...   1g of O_2 reacts to give $\frac{44}{32}$ = 1.375g of CO_2

So...   8.27g of O_2 will produce 8.27 × 1.375 = **11.37g of CO_2**

Calculating Volumes is pretty straightforward
Cover the page and have a go at this to make sure you've got it.
Find the volume of 2.5g of methane gas, CH_4, (at room T and P).

Electrolysis Calculations

The important bit here is to get the balanced half equations, because they determine <u>the relative amounts</u> of the two substances produced at the two electrodes. After that it's all the same as before, working out masses using M_r values, and volumes using the "M_r of any gas = 24 litres" rule.

The Three Steps for Electrolysis Calculations

1) Write down the <u>two balanced half equations</u> (i.e. match the number of electrons).

2) Write down the <u>balanced formulae</u> for the two products obtained from the two electrodes.

3) <u>Write in the M_r values</u> underneath each and carry on as for previous calculations.

Example

In the electrolysis of sodium chloride, hydrogen gas is released at the cathode and chlorine gas is released at the anode. If 0.25g of hydrogen gas is collected at thecathode, find the volume of chlorine released.

1) Balanced <u>half equations</u>:

$$2H^+ + 2e^- \rightarrow H_2 \quad \text{(2x1 because it's } H_2 \text{ not just H)}$$
$$2Cl^- - 2e^- \rightarrow Cl_2$$

(2x35.5 because it's Cl_2 not just Cl)

2) <u>Balanced formulae of products</u>:

(as obtained from the balanced half equations)

3) <u>Write in M_r values</u>:

...and carry on as usual

$$
\begin{array}{ccc}
 & H_2 & Cl_2 \\
\div 2 & 2 & 71 & \div 2 \\
 & 1 \dots\dots\dots & 35.5 \\
\times 0.25 & 0.25 \dots\dots & 8.875 & \times 0.25 \\
 & & = \underline{8.875\,g}
\end{array}
$$

So 0.25g of hydrogen will yield 8.875g of chlorine gas. Now we need this as a volume, so we use the good old "<u>mass to volume converting formula</u>":

$$\text{Volume} = \frac{\text{MASS}}{24,000} \; M_r \Rightarrow \text{Volume} = \frac{\text{MASS} \times 24,000}{M_r} = \frac{8.875 \times 24,000}{71} = \underline{3000\,\text{cm}^3}$$

And there it is. Just the same old stuff every time — balanced formulae, fill in M_r values, "divide and times" on both sides, and then use the "24,000 rule" to find the volume.

Calculation of A_r from % abundances of Isotopes

Some elements, like chlorine, have A_r values which are not whole numbers. This is because there are <u>two stable isotopes</u> of chlorine, ^{35}Cl and ^{37}Cl, and the mixture of the two gives an average A_r of 35.5. There is a simple formula for working out the overall A_r from the percentage abundances of two different isotopes:

$$\textbf{Overall } A_r = [(A_1 \times \%_{(1)}) + (A_2 \times \%_{(2)})] \div 100$$

Example

If chlorine consists of 76% ^{35}Cl and 24% ^{37}Cl, then the overall value for A_r is:

Overall $A_r = [(35 \times 76) + (37 \times 24)] \div 100 = 35.48 = \underline{35.5}$ to 1 d.p.

Keep your balance and pay attention

With electrolysis calculations the main tricky bit is getting the balanced half equations right. Make sure you have the <u>same</u> number of electrons in both half equations.

Warm-Up and Worked Exam Questions

Take a deep breath and go through these warm-up questions one by one.

Warm-up Questions

1) If 16g of CH_4 react to give 44g of CO_2, how much CO_2 would be produced from 50g of CH_4?
2) Calculate the volume of 3.5g of $O_{2(g)}$ (at room temperature/pressure).
3) The balanced equation for the reaction between sodium and chlorine is $2Na + Cl_2 \rightarrow 2NaCl$
   a) Calculate the mass of sodium chloride that will be produced when 50g of sodium react with chlorine gas.
   b) Calculate the mass of chlorine that is needed to react with the 50g of sodium.
   c) Find the volume of chlorine gas that reacts.
4) Find the mass of 7.5 dm³ of oxygen gas at room temperature/pressure (remember, 1dm³ = 1 litre).

Worked Exam Questions

There's no better preparation for exam questions than doing, err practice exam questions. Hang on, what's this I see...

1 One of the main ores of copper contains copper sulphide. Copper can be extracted easily from it:

$$CuS \rightarrow Cu + S$$

a) Calculate the mass of copper that will be produced from 192g of copper sulphide.

M_r of CuS = 96 M_r of Cu = 64   Work out the M_r of each.

Then divide for one...

1g CuS reacts to give $\frac{64}{96}$ = 0.67g Cu

...times for all. $0.67 \times 192 = 128g$ of Cu

(2 marks)

b) The sulphur produced in the reaction in part a) went on to react with oxygen to form sulphur dioxide:

$$S + O_2 \rightarrow SO_2$$

i) Calculate the mass of sulphur produced in part a).

192 - 128 = 64g of Sulphur

(1 mark)

ii) Calculate the volume of sulphur dioxide produced at room temperature/pressure.

M_r of S = 32 M_r of SO_2 = 64

1g S gives $\frac{64}{32}$ = 2g of SO_2 so mass of SO_2 = $2 \times 64 = 128g$

$volume = \frac{mass}{M_r} \times 24,000$

$\frac{128}{64} \times 24,000 = 48,000$ cm³ or 48 litres of SO_2

(3 marks)

Worked Exam Questions

Worked Exam Questions

2 When calcium carbonate decomposes, it forms calcium oxide and carbon dioxide.

a) Write the balanced equation for this reaction.

$$CaCO_3 \rightarrow CaO + CO_2$$

(1 mark)

b) What mass of calcium carbonate would produce 168g of calcium oxide?

$$M_r \text{ of } CaCO_3 = 40 + 12 + (16 \times 3) = 100 \qquad M_r \text{ of } CaO = 40 + 16 = 56$$

This time we're looking for the mass of reactant rather than product so... $\frac{100}{56} = 1.79g$ of $CaCO_3$ to produce 1g of CaO

...rather than $\frac{56}{100}$

$$1.79 \times 168 = 300g \text{ of } CaCO_3$$

(2 marks)

c) Calculate the volume of carbon dioxide that will be formed in the reaction.

$$CaO = 56 \qquad CO_2 = 12 + (16 \times 2) = 44$$

1g CaO produces... $\frac{44}{56} = 0.79g$ of CO_2 so... $0.79 \times 168 = 132g$ of CO_2 Work out the mass first.

Then apply the formula for finding the volume.

$$Vol\ CO_2 = \frac{132}{44} \times 24,000 = 72,000 \text{ cm}^3 \text{ or } 72 \text{ litres}$$

(3 marks)

3 Aluminium is extracted from aluminium oxide by electrolysis.

a) Write balanced half equations for the reactions happening at the cathode and anode.

At the cathode: $4Al^{3+} + 12e^- \rightarrow 4Al$ Remember to balance the electrons.

At the anode: $6O^{2-} \rightarrow 3O_2 + 12e^-$

(2 marks)

b) Find the mass of aluminium produced if 24g of oxygen is liberated in the reaction.

$$4Al \quad : \quad 3O_2$$

The half equations show that 4Al atoms are produced for every $3O_2$ molecules.

Work out the M_r of each... $4 \times 27 = 108 \qquad 3 \times (2 \times 16) = 96$

...then carry on as usual. $\frac{108}{96} = 1.125g \quad : \quad 1g$

$$1.125 \times 24 = 27g \text{ of } Al$$

(2 marks)

Exam Questions

1 A geologist discovers a sample of neon gas trapped in a large rock.
If the neon consists of 90% ^{20}Ne and 10% ^{22}Ne, calculate the overall (average) atomic mass, A_r, for this sample of neon.

....

....

....

(2 marks)

2 Iron is extracted by reducing iron(III) oxide in the presence of carbon monoxide.

a) Write the balanced equation for the reduction of iron(III) oxide by carbon monoxide to produce iron.

....

(2 marks)

b) Calculate the mass of iron produced from 80g of iron(III) oxide.

....

....

....

(2 marks)

c) What volume of carbon dioxide, would also be formed in the reaction at r.t.p.?

....

....

....

....

(3 marks)

d) Give two uses for iron.

....

....

(2 marks)

Exam Questions

3   Ammonia is used to make fertilisers.
     In the industrial production of ammonia, hydrogen and nitrogen react together.

   a)   Write the balanced equation for the reaction.

        ....
        (2 marks)

   b)   If 85g of ammonia are produced, what volume of hydrogen gas (measured at room
        temperature and pressure) must have reacted with nitrogen?

        ....

        ....

        ....

        ....
        (3 marks)

4   Many metal articles are plated with silver to make them more attractive. For example,
     teaspoons are often silver-plated. A spoon is used as the cathode in the electrolysis of a
     solution of silver nitrate. Silver coats the spoon and oxygen is liberated at the anode.

   a)   Write the balanced half equations for the reactions.

        ....

        ....
        (2 marks)

   b)   Each teaspoon has a surface area of $5.5cm^2$. The manufacturer wants to coat each spoon
        with a uniform layer of silver of 0.2g per cm^2. Calculate the mass of silver needed for
        each spoon.

        ....
        (1 mark)

   c)   Find the volume of oxygen evolved during the plating of 25 teaspoons.

        ....

        ....

        ....

        ....
        (3 marks)

The Mole

The Mole is really confusing. I think it's the word that puts people off. It's very difficult to see the relevance of the word "mole" to different-sized piles of brightly-coloured powders.

"THE MOLE" is simply the name given to *a certain number*

Just like "a million" is this many: 1 000 000; or "a billion" is this many: 1 000 000 000, so "a mole" is this many: 602 300 000 000 000 000 000 000   or 6.023×10^{23}.

1) And that's all it is. Just a number. The burning question, of course, is why is it such a silly long one like that, and with a six at the front?

2) The answer is that when you get precisely that number of atoms or molecules, of any element or compound, then, conveniently, they weigh exactly the same number of grams as the Relative Atomic Mass, A_r (or the relative formula mass) of the element or compound.
   This is arranged on purpose of course, to make things easier.

> **One mole of atoms or molecules of any substance will have a mass in grams equal to the Relative Formula Mass (A_r or M_r) for that substance.**

Example

Carbon has an A_r of 12.	So one mole of carbon weighs exactly 12g
Iron has an A_r of 56.	So one mole of iron weighs exactly 56g
Nitrogen gas, N_2, has an M_r of 28 (2×14).	So one mole of N_2 weighs exactly 28g
Carbon dioxide, CO_2, has an M_r of 44.	So one mole of CO_2 weighs exactly 44g

This means that 12g of carbon, or 56g of iron, or 28g of N_2, or 44g of CO_2, all contain the same number of atoms or molecules, namely one mole or 6×10^{23} atoms or molecules.

Don't worry, it's only a unit

Don't get confused about moles. The mole is just a unit like grams or metres — the only difference is that moles are a measure of the numbers of atoms. And a mole always weighs the same as the molecular mass so it's really easy to remember.

The Mole

*Nice Easy Formula for finding the **Number of Moles** in a **given mass**:*

$$\text{NUMBER OF MOLES} = \frac{\text{Mass in g} \quad \text{(of element or compound)}}{M_r \quad \text{(of element or compound)}}$$

Example

How many moles are there in 42g of carbon?

No. of moles $= \text{Mass (g)} / M_r = 42/12 = \underline{3.5 \text{ moles}}$

"Relative Formula Mass" is also "Molar Mass"

1) We've been very happy using the Relative Formula Mass, M_r, all through the calculations.

2) In fact, that was already using the idea of moles because M_r is actually the mass of one mole in g, or as we sometimes call it, the <u>molar mass</u>.

3) It follows that <u>the volume of one mole of any gas</u> will be <u>24 litres</u> — <u>the molar volume</u>.

A "One Molar Solution" Contains "One Mole per Litre"

This is pretty easy. So a 2M (two molar) solution of NaOH contains 2 moles of NaOH per litre of solution.
You need to know how many moles there'll be in a given volume of solution:

$$\text{NUMBER OF MOLES} = \underline{\text{Volume} \text{ in Litres}} \times \underline{\text{Molarity} \text{ of solution}}$$

Example

How many moles in 185cm³ of a 2M solution?

<u>ANS:</u> $0.185 \times 2 = \underline{0.37 \text{ moles}}$

Moles — a suitably silly name for such a confusing idea...

It's possible to do all the calculations on the previous pages without ever talking about moles. You just concentrate on M_r and A_r all the time instead. In fact M_r and A_r represent moles anyway, but <u>I</u> think it's less confusing if moles aren't mentioned at all. Ask "Teach" how much of these pages you should learn.

Warm-Up and Worked Exam Questions

Have a go at these warm-up questions to make sure you've got moles 'in the bag'.

Warm-up Questions

1) How many atoms are there in 24g of magnesium?
2) What is the mass of 6.023×10^{23} atoms of fluorine?
3) What is the mass of 6.023×10^{23} molecules of sulphur trioxide SO_3?
4) What is the mass of one mole of sulphur trioxide SO_3?
5) The relative formula mass of sodium fluoride (NaF) is 42. What is its molar mass?

Worked Exam Question

Exam questions are the best way to practise what you've learnt. After all they're exactly what you'll have to do on the big day — so work through this worked example very carefully.

1 Milk of Magnesia or magnesium hydroxide is used to neutralise excess acidity in the stomach. It is called antacid.
The formula of magnesium hydroxide is $Mg(OH)_2$

a) How many moles are in 174g of magnesium hydroxide.

$24 + 2(16 + 1) = 58$ = the relative molecular mass of $Mg(OH)_2$

1 mole of $Mg(OH)_2$ = 58g $\dfrac{174}{58} = 3 \text{ moles}$

(2 marks)

b) Only 0.009g of magnesium hydroxide can be dissolved in one litre ($1000cm^3$) of water. This compares with 420g of sodium hydroxide.

i) How many grams of sodium hydroxide are needed to make $500cm^3$ of 0.1M solution?

$23 + 16 + 1 = 40$ = the relative molecular mass of NaOH

$500cm^3$ = 0.5 litres → $0.5 \times 0.1 = 0.05 \text{ moles}$ = number of moles of NaOH needed
We want a 0.1 M solution

$0.05 \times 40 = 2g$

(2 marks)

ii) Find the molar concentration of a saturated solution of magnesium hydroxide.

$Mg(OH)_2 = 24 + 2(16 + 1) = 58$

$\dfrac{0.009}{58} = 0.00016 \text{ M}$

(2 marks)

Exam Questions

1   'Tummy Tamer' is an antacid which contains sodium hydrogencarbonate $NaHCO_3$.

The balanced chemical equation for the reaction of $NaHCO_3$ and hydrochloric acid is:

$$NaHCO_3 + HCl \rightarrow NaCl + H_2O + CO_2$$

a)   Using this equation calculate the following quantities:

   i)   The mass of sodium hydrogencarbonate which will react with 36.5g of hydrochloric acid?

   ....

   ....

   (2 marks)

   ii)   The number of moles of sodium chloride produced from 36.5g of hydrochloric acid.

   ....

   ....

   (1 mark)

   iii)   The volume of carbon dioxide gas produced in the reaction (at room temperature/ pressure). (The volume of one mole of any gas is 24 litres)

   ....

   ....

   (2 marks)

b)   How many molecules are there in 4000cm^3 of carbon dioxide?

....

....

....

(2 marks)

c)   If 'Tummy Tamer' was used to neutralise a 0.4M solution of HCl, what volume of the acid would be needed to produce the same quantity of NaCl as in part a)?

....

....

....

(3 marks)

Revision Summary

Why bother doing easy questions? These monsters find out what you really know, and worse, what you really don't. I know, it's kind of scary, but if you want to get anywhere in life you've got to face up to a bit of hardship. That's just the way it is. Take a few deep breaths and then try these:

1) Describe each of these nine types of chemical change: Thermal decomposition, neutralisation, displacement, precipitation, oxidation, reduction, exothermic, endothermic, reversible.

2) Give three rules for balancing equations. Balance these and put the state symbols in:
   a) $CaCO_3 + HCl \rightarrow CaCl_2 + H_2O + CO_2$
   b) $Ca + H_2O \rightarrow Ca(OH)_2 + H_2$
   c) $H_2SO_4 + KOH \rightarrow K_2SO_4 + H_2O$
   d) $Fe_2O_3 + H_2 \rightarrow Fe + H_2O$
   e) propane + oxygen $\rightarrow$ carbon dioxide + water

3) What is electrolysis? What is needed for electrolysis to take place?

4) Draw a diagram showing the electrolysis of NaCl solution. What are cations and anions?

5) What is the rule for balancing half equations?

6) What are A_r and M_r?

7) What is the relationship between A_r and the number of protons and neutrons in the atom?

8) Find A_r or M_r for these (use the periodic table inside the front cover):
   a) Ca   b) Ag   c) CO_2 d) $MgCO_3$ e) Na_2CO_3 f) ZnO   g) KOH
   h) NH_3 i) butane   j) sodium chloride   k) Iron(II) chloride

9) What is the formula for calculating the percentage mass of an element in a compound?
   a) Calculate the percentage mass of oxygen in magnesium oxide, MgO
   b) Calculate the percentage mass of carbon in   i) $CaCO_3$ ii) CO_2 iii) Methane
   c) Calculate the percentage mass of metal in these oxides: i) Na_2O ii) Fe_2O_3 iii) Al_2O_3

10) What is meant by an empirical formula?

11) List the five steps of the method for finding an empirical formula (EF) from masses or %.

12) Work these out (using the periodic table):
   a) Find the EF for the iron oxide formed when 45.1g of iron reacts with 19.3g of oxygen.
   b) Find the EF for the compound formed when 227g of calcium reacts with 216g of fluorine.
   c) Find the EF for when 208.4g of carbon reacts with 41.7g of hydrogen.
   d) Find the EF when 21.9g of magnesium, 29.3g of sulphur and 58.4g of oxygen react.

13) Write down the three steps of the method for calculating reacting masses.
   a) What mass of magnesium oxide is produced when 112.1g of magnesium burns in air?
   b) What mass of sodium is needed to produce 108.2g of sodium oxide?
   c) What mass of carbon will react with hydrogen to produce 24.6g of propane?

14) What mass of gas occupies 24 litres at room temperature and pressure?

15) Write down the formula for calculating the volume of a known mass of gas (at room T & P).
   a) What is the volume of 56.0g of nitrogen at room T & P?
   b) Find the volume of carbon dioxide produced when 5.6g of carbon is completely burned.
   c) What volume of oxygen will react with 25.0g of hydrogen to produce water?
   d) What is the mass of 5.4 litres of nitrogen gas?
   e) What mass of carbon dioxide is produced when 4.7 litres of oxygen reacts with carbon?

16) Write down the three steps for electrolysis calculations.
   a) In the electrolysis of molten NaCl, find the mass of Cl_2 released if 3.4g of sodium are collected.
   b) In the electrolysis of copper(II) chloride, what volume of chlorine gas would be produced for every 100g of copper obtained?

17) What is a mole? Why is it that precise number?

18) How much does one mole of any compound weigh? What is meant by molar mass?

19) What is the molar volume of a gas? What is meant by a "1 molar" or "2 molar" solution?

The Evolution of the Atmosphere

The present composition of the atmosphere is: <u>78% nitrogen</u>, <u>21% oxygen</u>, <u>0.04% CO_2</u> (= 99.04%). The remaining 1% is made up of noble gases (mainly argon). In addition there can be a lot of water vapour. But the atmosphere wasn't <u>always</u> like this. Here's how the first 4.5 billion years have gone:

Phase 1 — *Volcanoes* gave out *Steam, CO_2, NH_3 and CH_4*

1) The Earth's surface was originally <u>molten</u> for many millions of years. Any atmosphere <u>boiled away</u>.
2) Eventually it cooled and a <u>thin crust</u> formed but <u>volcanoes</u> kept erupting.
3) They belched out mostly <u>carbon dioxide</u> and some <u>steam</u>.
4) The early atmosphere was <u>mostly CO_2</u>, <u>carbon monoxide</u> and <u>water vapour</u>.
5) There was virtually <u>no oxygen</u>.
6) The water vapour <u>condensed</u> to form the <u>oceans</u>.

Phase 2 — *Green Plants Evolved* and produced *Oxygen*

1) <u>Green plants</u> evolved over most of the Earth.
2) They were quite happy in the <u>CO_2 atmosphere</u>.
3) A lot of the early CO_2 <u>dissolved</u> into the oceans.
4) But the <u>green plants</u> steadily <u>removed CO_2</u> and <u>produced O_2</u> by photosynthesis.
5) Much of the CO_2 from the air thus became <u>locked up</u> in <u>fossil fuels</u> and <u>sedimentary rocks</u>.
6) <u>Methane</u> and <u>ammonia</u> reacted with the <u>oxygen</u>, releasing <u>nitrogen gas</u>.
7) Ammonia was also converted into <u>nitrates</u> by nitrifying bacteria.
8) <u>Nitrogen gas</u> was also released by <u>living organisms</u> like denitrifying bacteria.

Phase 3 — *Ozone Layer* allows *Evolution* of *Complex* Animals

1) The build-up of <u>oxygen</u> in the atmosphere <u>killed off</u> early organisms that couldn't tolerate it.
2) It also enabled the <u>evolution</u> of more <u>complex</u> organisms that <u>made use</u> of the oxygen.
3) The oxygen also created the <u>ozone layer</u> (O_3), which <u>blocked</u> harmful rays from the sun and <u>enabled</u> even <u>more complex</u> organisms to evolve.
4) There is virtually <u>no CO_2</u> left now.

---

4 million years ago was a whole other world

It's pretty amazing how much the atmosphere has changed. It makes our present day obsession about the CO_2 going up from 0.03% to 0.04% seem a bit ridiculous, doesn't it.

Today's Atmosphere

The atmosphere we have today is <u>just right</u>.
It has <u>gradually evolved</u> over billions of years and <u>we</u> have evolved with it. All very slowly.
It's been about the same for the past 200 million years. We worry that we're changing it <u>for the worse</u> by releasing various gases from <u>industrial activity</u>. There are three main worries:
The <u>Greenhouse Effect</u>, The <u>Ozone Layer</u> and <u>Acid Rain</u>. These are described later in the section. (And in more detail in the Biology Book.)

Composition of Today's Atmosphere

Present composition of the atmosphere:

78%	Nitrogen	} (Often written as 79%
1%	Argon	Nitrogen for simplicity.)
21%	Oxygen	
0.04%	Carbon dioxide	

(That comes to over 100% because the first three are rounded up very slightly)

Also :
1) Varying amounts of WATER VAPOUR.
2) And other noble gases in very small amounts.

A Simple Experiment to find the % of Oxygen in the Air

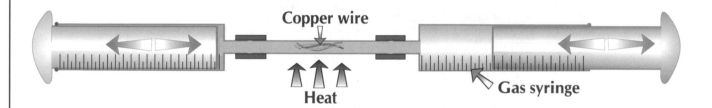

Copper wire

Heat

Gas syringe

Method

1) Measure the <u>initial volume</u> of air, then push it <u>back and forth</u> over the <u>heated copper</u>.

2) The copper <u>takes out</u> the oxygen and produces <u>black copper oxide</u>.

3) When <u>no more</u> copper is turning black, let it <u>cool</u> and measure the <u>amount</u> of air left.

4) As a <u>check</u>, <u>heat</u> the copper <u>again</u> for a while, cool and <u>measure the volume again</u>.

5) Then <u>calculate</u>:

$$\text{Percentage of oxygen} = \frac{\text{Change in volume}}{\text{Original volume}} \times 100$$

Oxygen may be important, but it's not the only gas in here

This is all pretty simple. It's the kind of stuff that some of you might have thought doesn't need to be learnt. You should know better than that — it's here because you need it.

The Carbon Cycle

The *Oceans* Hold a lot of *Carbon*

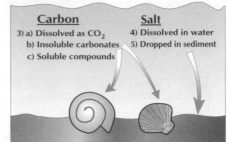

1) The <u>Oceans</u> were formed by <u>condensation</u> of the <u>steam</u> in the early atmosphere.

2) They then started <u>absorbing</u> the CO_2 from the atmosphere.

3) They now contain <u>a large amount of carbon</u> in <u>three</u> main forms:
   a) <u>Carbon dioxide dissolved</u> in the water
   b) <u>Insoluble carbonates</u> like calcium carbonate
      (e.g. shells, which form sediment and then limestone)
   c) <u>Soluble compounds</u> like hydrogen-carbonates of Ca, Mg

4) The Oceans are gradually getting <u>saltier</u>.

5) Some salts are <u>removed</u> by various <u>chemical</u> and <u>biological</u> processes and dropped as <u>sediment</u>.

6) This eventually turns into <u>rocks</u> and these eventually <u>wash back in again</u> millions of years later.

Carbon Cycle

The <u>Carbon Cycle</u> shown below is a summary of how carbon passes through various forms and is constantly <u>recycled.</u> It's not as bad as it looks. Well, not once you <u>know it all</u> anyway.

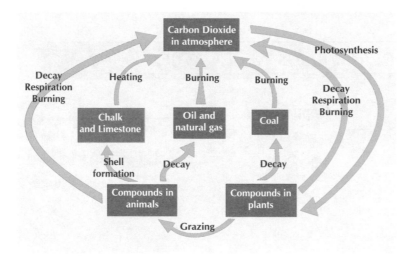

1) There's <u>a better version</u> of the carbon cycle given in the <u>Biology Book</u>.

2) There are <u>many</u> different ways of representing the carbon cycle, but in the end <u>they all show the same things happening</u>. If you <u>properly understand</u> one diagram you should be able to deal with any other, even if it looks totally different. Apart from the pretty colours this is a <u>standard syllabus version</u>. That means it's like the one you'll probably get in your Exam.

3) This diagram just shows all the information, but <u>without</u> trying to make it <u>clear</u>!
   Look at how this relates to the one in the Biology Book, which does try to make it clear.

4) This one also contains <u>chalk</u> and <u>limestone</u> which were left out of the biology one.

5) You should learn about all these processes elsewhere. This diagram is just a <u>summary</u> of them.

6) In the Exam they could give you this diagram with <u>labels missing</u> and you'd have to <u>explain</u> or <u>describe</u> the missing process, so it's <u>pretty important</u> that you understand the <u>whole</u> thing and know about each process.

The Carbon Cycle is a multi-purpose beauty
The Carbon Cycle is one of the great highlights of the Double Science Syllabus — and handy for your Biology Exam as well as for Chemistry.

Man-made Atmospheric Problems

There are three man-made atmospheric problems on the syllabus.
Don't confuse them, they're all totally separate. (The Biology Book has more details on these.)

1) Acid Rain is caused by Sulphur Dioxide and Nitrogen Oxides

1) When fossil fuels are burned they release mostly CO_2 (which causes the Greenhouse Effect).

2) But they also release two other harmful gases, sulphur dioxide and various nitrogen oxides.

3) The sulphur dioxide, SO_2, comes from sulphur impurities in the fossil fuels.

4) However, the nitrogen oxides are created from a reaction between the nitrogen and oxygen in the air, caused by the heat of the burning.

5) When these gases mix with clouds they form dilute sulphuric acid and dilute nitric acid.

6) This then falls as acid rain.

7) Cars and power stations are the main causes of acid rain.

Acid Rain Kills Fish, Trees and Statues

1) Acid rain causes lakes to become acidic and many plants and animals die as a result.

2) Acid rain kills trees and damages limestone buildings and ruins stone statues.

Bad for the trees but good for you...

...because acid rain can get you some nice easy marks in the Exam. Make sure you remember the names of the two gases which cause acid rain, and where they come from.

Man-made Atmospheric Problems

2) The **Greenhouse Effect** is caused by CO_2 trapping heat

1) The Greenhouse Effect is causing the Earth to warm up very slowly.

2) It's caused mainly by a rise in the level of CO_2 in the atmosphere due to the burning of massive amounts of fossil fuels in the last two hundred years or so.

3) The carbon dioxide (and a few other gases) trap the heat that reaches Earth from the sun.

4) This will cause a rise in temperature which is then likely to cause changes in climate and weather patterns all over the world and possible flooding due to the polar ice caps melting.

5) The level of CO_2 in the atmosphere has gone up by about 20%, and will continue to rise as long as we keep burning fossil fuels, as the graph clearly shows.

6) Deforestation is not helping either.

7) The increased concentration of CO_2 in the atmosphere means the ocean surfaces absorb a bit more CO_2, but not enough to stop the rising levels.

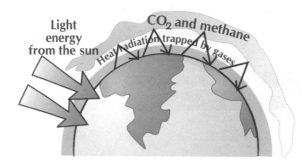

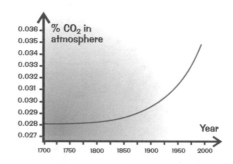

3) **CFCs** (from aerosols) Cause **The Hole** in **The Ozone Layer**

1) Ozone is a molecule made of three oxygen atoms, O_3.

2) There's a layer of ozone high up in the atmosphere.

3) It absorbs harmful UV rays from the sun.

4) CFC gases react with ozone molecules and break them up.

5) This thinning of the ozone layer allows harmful UV rays to reach the surface of the Earth.

But this has nothing whatever to do with the Greenhouse Effect or acid rain. Don't mix them up:

> CFCs = OZONE LAYER
>
> CO_2 = GREENHOUSE EFFECT
>
> SO_2 and NO_X = ACID RAIN

The Greenhouse Effect has nothing to do with Ozone or acid rain

Make sure you know your atmospheric problems inside out. They love to ask questions on them, and it's easy to muddle them up. You know the drill: learn, cover, scribble, check... learn...

Warm-Up and Worked Exam Questions

I know that you'll be champing at the bit to get into the exam questions, but these basic warm-up questions are invaluable to get the basic facts straight first.

Warm-up Questions

1) Write down three of the main gases in the Earth's original atmosphere.
2) Describe 2 processes that caused nitrogen to appear in the atmosphere.
3) Write down the name of the main noble gas in the air today.
4) Describe 2 processes that remove carbon dioxide from the atmosphere.
5) Describe 3 effects of acid rain.
6) Write down the formula for ozone.

Worked Exam Questions

Exam questions don't vary that wildly and the basic format is the same.
So you'd be mad not to spend a bit of time learning a model answer, wouldn't you...

1   Describe how the following processes affect levels of carbon dioxide and oxygen in the atmosphere.

a)   Photosynthesis

Photosynthesis by plants removes CO_2 from the atmosphere, and releases oxygen.

(1 mark)

b)   Respiration

Respiration by plants and animals releases CO_2 into the atmosphere, and removes oxygen.

(1 mark)

It's a good idea to write down what causes these processes to take place.

c)   Combustion

Combustion of fossil fuels releases CO_2 into the atmosphere and removes oxygen.

(1 mark)

2   The ozone layer is very important.
Describe and explain its importance and how it is being damaged.

The ozone layer is important because it absorbs harmful UV rays from the sun. It is being damaged by CFCs, which react with and break up ozone molecules.

(3 marks)

Exam Questions

1   The diagram below shows part of the carbon cycle.
     Identify the four missing processes to show how carbon passes through the environment.

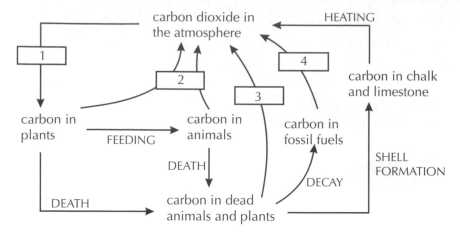

1 = ...

2 = ...

3 = ...

4 = ...

(4 marks)

2   The amounts of oxygen and carbon dioxide in the atmosphere have changed over the last
     4.5 billion years.

   a)   Describe what happened to the amount of oxygen and explain how the change came
        about.

        ....

        ....
                                                                                                    (2 marks)

   b)   Describe what happened to the amount of carbon dioxide and explain how the change
        came about.

        ....

        ....
                                                                                                    (2 marks)

Exam Questions

3   A student carried out an experiment to determine the amount of oxygen present in the atmosphere today. She used the apparatus below and, over a short period of time, all the oxygen reacted with the copper turnings. The diagram shows the end of the experiment.

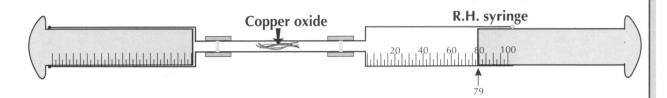

Copper oxide   R.H. syringe

20   40   60   80   100
79

a)   The syringe originally contained 100cm^3 of air. Calculate the volume of air that reacted.

....
(1 mark)

b)   Write down the percentage of oxygen in the air according to the student's result.

....
(1 mark)

c)   Fill in the table below to show the percentages of the 4 main gases in the air today.

Gas	Percentage
Nitrogen	78%
	0.04%
	1%
	21%

(3 marks)

4   Atmospheric pollution is causing a number of environmental problems.

a)   State 2 gases that lead to the formation of acid rain.

....
....
(2 marks)

b)   What is the main source of these gases.

....
(1 mark)

The Three Different Types of Rocks

Rocks shouldn't be confusing. There are <u>three</u> different types: <u>sedimentary</u>, <u>metamorphic</u> and <u>igneous</u>. Over <u>millions of years</u> they <u>change from one into another</u>. This is called the <u>Rock Cycle</u>.

The Rock Cycle

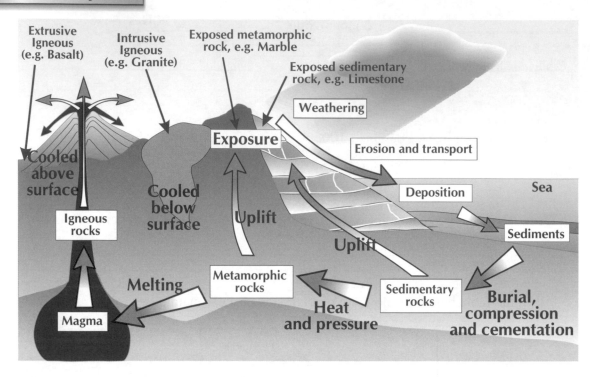

The Rocks Change from One to Another in a Slow Cycle

1) Particles get <u>washed to the sea</u> and settle as <u>sediment</u>.

2) Over <u>millions of years</u> these sediments get <u>crushed</u> into <u>SEDIMENTARY</u> rocks (hence the name).

3) At first they get <u>buried</u>, but they can either <u>rise to the surface</u> again to be discovered, or they can <u>descend</u> into the <u>heat</u> and <u>pressure</u> below.

4) If they <u>descend</u>, the heat and increased pressure <u>completely alter</u> the <u>structure</u> of the rock and they then become <u>METAMORPHIC ROCKS</u> (as in "metamorphosis" or "change" — another good name).

5) These <u>metamorphic rocks</u> can either <u>rise to the surface</u> to be discovered by an enthusiastic geologist or else descend <u>still further</u> into the <u>fiery abyss</u> of the Earth's raging inferno (the mantle) where they will partially <u>melt</u> and become <u>magma</u>.

6) When <u>magma</u> reaches the surface it <u>cools</u> and <u>sets</u> and is then called <u>IGNEOUS ROCK</u> ("igneous" as in "ignite" or "fire" — another great name).

7) There are actually <u>two types</u> of igneous rock:
   a) <u>EXTRUSIVE</u> when it comes <u>straight out</u> of the surface from a <u>volcano</u> ("Ex-" as in "Exit").
   b) <u>INTRUSIVE</u> when it just sets as a big lump <u>below</u> the surface ("In-" as in "inside").
      (I have to say — whoever invented these names deserves a medal.)

8) When any of these rocks reach the <u>surface</u>, then <u>weathering</u> begins and they gradually get <u>worn down</u> and carried off <u>to the sea</u> and the whole cycle <u>starts over again</u>...

Rocks are a mystery — no, no, it's sedimentary my Dear Watson...

Can you think of anything better than going on a family holiday to Cornwall, gazing at the cliffs and marvelling at the different types of rock? Of course you can, but it still beats more tedious equations.

Sedimentary Rocks

Three steps in the Formation of Sedimentary Rocks

1) <u>Sedimentary rocks</u> are formed from <u>layers of sediment</u> laid down in <u>lakes</u> or <u>seas</u>.

2) Over <u>millions of years</u> the layers get <u>buried</u> under more layers and the <u>weight</u> pressing down <u>squeezes out</u> the water.

3) Fluids flowing through the pores deposit natural mineral cement.

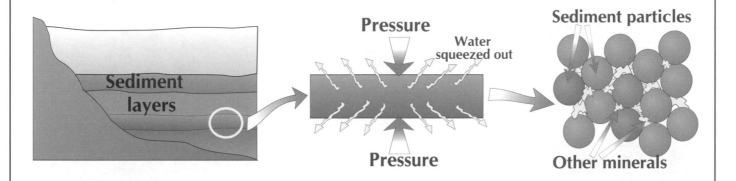

The Deepest Rocks are usually the Oldest Rocks

1) Because of the order they were laid down in, the <u>deepest rocks</u> are usually the <u>oldest rocks</u>.

2) If you get a group of rocks in a nice banded pattern, with another rock <u>cutting across</u> them, the rock that cuts across must have been laid down <u>after</u> the other layers — so it'll be <u>younger</u>.

Fossils are mainly found in Sedimentary Rocks

1) Only <u>sedimentary</u> rocks and some metamorphic rocks (like marble) that were once sediments contain <u>fossils</u>. The heat and presssure of <u>metamorphism</u> eventually destroys fossils.

2) Sedimentary rocks have only been <u>gently crushed</u> for a few million years. No big deal, so the <u>fossils survive</u>. All sedimentary rocks are likely to contain fossils.

3) Fossils are a very useful way of <u>identifying rocks</u> as being of the <u>same age</u>.

4) This is because the plants/animals that leave behind their fossilised remains <u>change</u> (due to evolution) as the <u>ages pass</u>.

5) This means that if two rocks have the <u>same fossils</u> they must be from the <u>same age</u>.

6) However, if the fossils in two rocks are <u>different</u>, it proves <u>nothing</u> don't forget.

Not much to look at but pretty interesting all the same...

Sedimentary rocks are rather the ugly sisters of the rock world. They tend to be coarse, dull, and kind of gritty. You'll see what I mean when you examine the four specimens on the next page...

Sedimentary Rocks

The **Four** Main **Sedimentary Rocks**:

Sedimentary rocks tend to <u>look similar</u> to the <u>original sediments</u> from which they formed. After all, <u>very little</u> has happened other than them <u>squashing together</u> and cementing the grains together.

1) *Sandstone*

This is formed from <u>sand</u> of course. And it looks like it too. Sandstone just looks like <u>sand particles</u> all stuck <u>very firmly</u> together. There's <u>red</u> sandstone and <u>yellow</u> sandstone, which are commonly used for <u>buildings</u>.

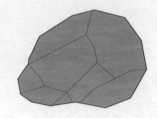

2) *Limestone*

This formed from <u>seashells</u>. It's mostly <u>calcium carbonate</u> and <u>grey/white</u> in colour.
The original <u>shells</u> are mostly <u>crushed</u> but there are still quite a few <u>fossilised shells</u> to be found in <u>limestone</u>.

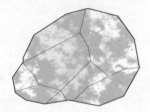

3) *Mudstone or shale*

This was formed from <u>mud</u>, which basically means <u>finer particles than sand</u>. It's often <u>dark grey</u> and tends to <u>split</u> into the <u>original layers</u> very easily.

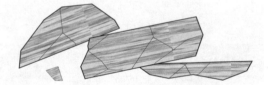

4) *Conglomerates*

These look like a sort of crude <u>concrete</u>, containing <u>pebbles</u> set into a <u>cement</u> of finer particles.

Know your sedimentary rocks

Quite a lot of facts here on sedimentary rocks. You've got to <u>learn</u> how they form, how to tell which are older, that they contain fossils, and also the names etc. of the four examples. Most importantly you need to be able to <u>describe</u> in words <u>what they all look like</u>. Make sure you learn the descriptions.

Metamorphic Rocks

Heat and Pressure over Thousands of Years

Metamorphic rocks are formed by the action of heat and pressure on existing (sedimentary) rocks over long periods of time. You know that the rocks are changed versions of other rocks because they have the same chemical compositions.

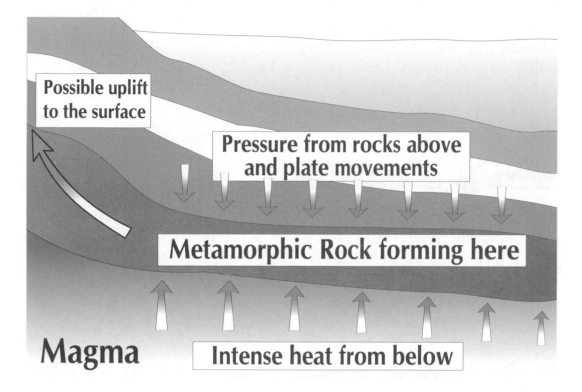

1) Earth movements can push all types of rock deep underground.

2) Here they are compressed and heated, and the mineral structure and texture may change.

3) So long as they don't actually melt they are classed as metamorphic rocks.

4) If they melt and turn to magma, they're gone. The magma may resurface as igneous rocks.

Don't get confused about metamorphic rocks

"Metamorph" (as in metamorphosis etc.) means to change form, and that's what metamorphic rocks are all about. It's pretty simple really. They're just old rocks that have got squidged and heated till they don't know what they are or where they're from. Poor things.

Metamorphic Rocks

Slate, Marble and Schist are **Metamorphic Rocks**

1) **Slate** is formed from **mudstone** or **clay**

1) As the mudstone gets heated and compressed its tiny plate-like particles align in the same direction.

2) This allows the resulting slate to be split along that direction into thin sheets which make ideal roofing material.

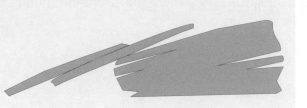

2) **Marble** is formed from **Limestone**

1) Very high temperature will break down the shells in limestone, which then reform as small crystals.

2) This gives marble a more even texture and makes it much harder.

3) It can be polished up and often has attractive patterning.

4) This makes it a great decorative stone.

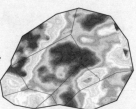

3) **Schist** is formed when **Mudstone** gets very **hot**

1) Mudstone will turn to slate only if there's plenty of pressure but not too much heat.

2) If mudstone gets really hot, new minerals like mica start to form and create layers.

3) This creates schist, a rock containing bands of interlocking crystals.

4) These layers of crystals are typical of a metamorphic rock.

5) Only steady heat and pressure will cause this to happen.

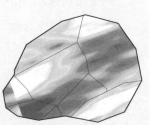

Metamorphic rocks are pretty

There's quite a lot of names accumulating now. Somehow, you've got to make sense of them in your head. Knowing what these rocks actually look like can be a real help.

Igneous Rocks

Igneous Rocks are formed from Fresh Magma

1) Igneous rocks form when molten magma pushes up into the crust or right through it.

2) Igneous rocks contain various different minerals in randomly arranged interlocking crystals.

3) There are two types of igneous rocks: EXTRUSIVE and INTRUSIVE:

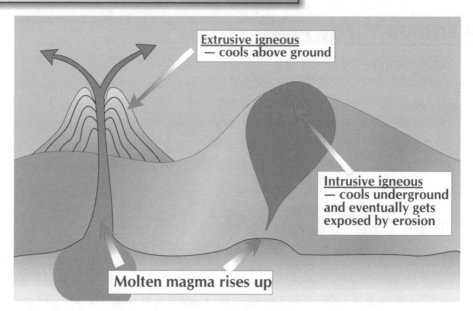

Extrusive igneous — cools above ground

Intrusive igneous — cools underground and eventually gets exposed by erosion

Molten magma rises up

INTRUSIVE igneous rocks cool SLOWLY with BIG crystals

GRANITE is an **intrusive** igneous rock with **big crystals**

1) Granite is formed underground where the magma cools down slowly.

2) Because it cools down slowly, it has big randomly arranged crystals .

3) Granite is a very hard and decorative stone ideal for steps and buildings.

EXTRUSIVE igneous rocks cool QUICKLY with SMALL crystals

BASALT is an **extrusive** igneous rock with **small crystals**

1) Basalt is formed on top of the Earth's crust after bursting out of a volcano.

2) This means it has relatively small crystals — because it cooled quickly.

Identifying Rocks in Exam Questions

A typical question will simply describe a rock and ask you to identify it. Make sure you learn the information on rocks well enough to work backwards, as it were, so that you can identify the type of rock from a description. Practise by doing these:

Rock A: Small crystals in layers.

Rock B: Contains fossils.

Rock C: Randomly arranged crystals of various types.

Rock D: Hard, smooth and with wavy layers of crystals.

Rock E: Large crystals. Very hard wearing.

Rock F: Sandy texture. Fairly soft.

Answers

A: metamorphic
B: sedimentary
C: igneous
D: metamorphic
E: igneous (granite)
F: sedimentary (sandstone)

Igneous Rocks are real cool — or they're magma...

It's important that you know what granite looks like. You should insist that you go on a field trip to the famous pink granite coast of Brittany. Two weeks should be enough to fully appreciate it.

Warm-Up and Worked Exam Questions

Learning facts and practising exam questions is the only recipe for success. That's what the questions on these pages are all about. All you have to do — is do them.

Warm-up Questions

1) A rock has been formed from seashells squashed together for millions of years. Name the rock and state which rock type it belongs to.

2) Describe the processes that affect clay as it changes into slate.

3) Write down the name for the molten rock from which igneous rocks are formed.

4) Describe one use for sandstone and one for marble.

5) Compare the processes by which intrusive and extrusive igneous rocks are formed.

Worked Exam Question

I'd like an exam question, and the answers written in — and a surprise. Two out of three's not bad.

1   The diagram below shows the cross-section through a volcano.

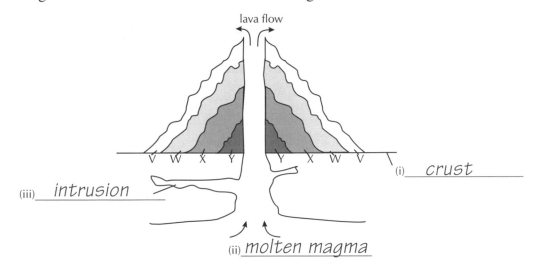

lava flow

(i) _crust_

(iii) _intrusion_

(ii) _molten magma_

a) Complete the missing labels.

(3 marks)

b) Name the type of rocks produced by volcanoes.

........._Igneous_...

(1 mark)

c) Which of the labelled rock layers is the youngest?

........._V_...

(1 mark)

d) Explain why the crystals in the middle of rock layer V are larger than the crystals on the surface of rock layer V.

Crystals in the middle of rock V are larger because they were

exposed to air for less time and therefore cooled more slowly.

(2 marks)

Exam Questions

1   The table below gives some information about six different rocks A, B, C, D, E and F.

Rock	Does it fizz in acid?	Is it hard?	Does it have crystals?	Are the crystals banded?	Does it have fossils?
A	no	yes	yes	yes	no
B	no	yes	yes, large ones	no	no
C	yes	no	no	-	yes
D	no	no	no	-	no
E	no	no	no	-	no
F	no	yes	yes, small ones	no	no

a)   Write down the letter of

   i)   a sedimentary rock

   ....

   ii)   a metamorphic rock

   ....

   iii)   an igneous rock

   ....

(3 marks)

b)   Write down the letter of a rock that might be

   i)   schist

   ....

   ii)   basalt

   ....

   iii)   sandstone

   ....

(3 marks)

c)   Rocks B and F have the same chemical composition. Suggest a possible reason for their different structures.

   ....

   ....

(2 marks)

Exam Questions

2   The diagram below shows part of the rock cycle. Fill in the missing boxes.

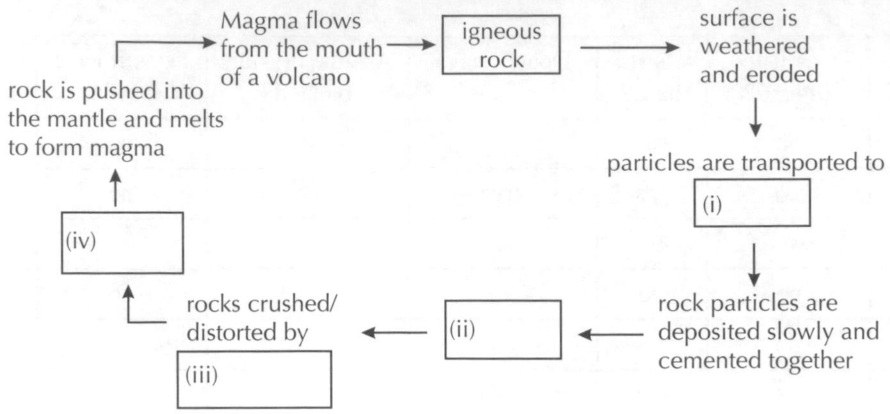

Magma flows from the mouth of a volcano → igneous rock → surface is weathered and eroded

rock is pushed into the mantle and melts to form magma

(iv)

rock is pushed into the mantle and melts to form magma ← (iv)

particles are transported to (i)

rocks crushed/ distorted by (iii) ← (ii) ← rock particles are deposited slowly and cemented together

(4 marks)

3   a)   Describe how sedimentary rocks are formed in the Earth's crust.

....

....

....

(3 marks)

b)   Describe the appearance of conglomerate rock.

....

....

(1 mark)

c)   A rock is very dark in colour and consists of bands of crystals.
Suggest the rock type to which it may belong.

....

(1 mark)

4   The diagrams below show typical rock micrograph slides of 2 different types of rock, P and Q.

FOSSIL

P   Q

a)   What types of rock are P and Q?

....

(2 marks)

b)   Describe how rock Q might be formed in the Earth's crust.

....

....

(2 marks)

Weathering and the Water Cycle

Weathering is the process of breaking rocks up

There are <u>three</u> distinct ways that rocks are <u>broken up</u> into small <u>fragments</u>:

A) *Physical weathering is caused by ice in cracks*

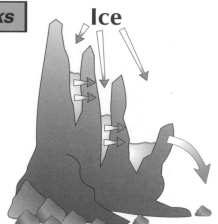

Ice

1) <u>Rain water</u> seeps into <u>cracks</u> in rocks and if the temperature drops <u>below freezing</u>, the water turns to <u>ice</u> and the <u>expansion</u> pushes the rocks apart.
2) This keeps happening <u>each time</u> the water <u>thaws and refreezes</u>.
3) After freezing and thawing many times bits <u>break off</u>.

B) *Chemical weathering is caused by acidic rain on limestone*

1) Remember, limestone is mainly <u>calcium carbonate</u> ($CaCO_3$), which will react with acid and dissolve away.
2) This isn't just "acid rain" caused by pollution. <u>Ordinary rain</u> is <u>weakly acidic</u> anyway, so it very gradually <u>dissolves</u> all <u>limestone</u>.

C) *Biological weathering is caused by plant roots in cracks*

<u>Plants</u> push their <u>roots</u> through cracks in rocks and as the roots <u>grow</u> they gradually <u>push the rocks apart</u>.

The weathering report for today — Ice, Acid and Plants

Of course, weathering is actually really exciting and interesting. Okay, so maybe that's going too far, but still, have you never wondered why some valleys are all untidy and covered in boulders? Well, it's because ice and trees and stuff are slowly chipping bits off the mountains.

Weathering and the Water Cycle

Erosion and *Transport*

1) <u>Erosion</u> is the <u>wearing away</u> and removal of exposed rocks by any means.
   It's different from weathering.

2) <u>Transport</u> is the process of <u>carrying away</u> the rock fragments — they can either <u>fall away</u> due to gravity, or be carried away <u>by rivers</u>. The rocks travelling down rivers get <u>worn down</u> as they go and they also wear away the <u>river bed</u> causing <u>river valleys</u>. The Grand Canyon is a grand example.

The *Water Cycle*

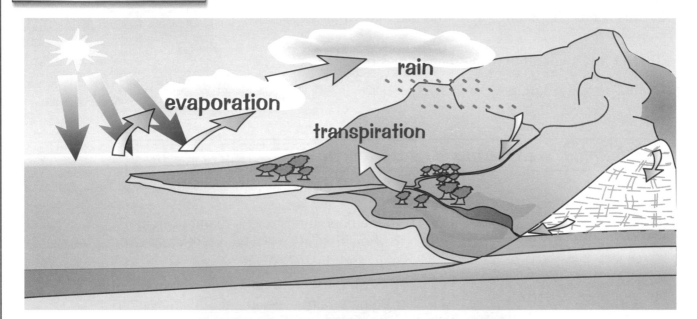

THIS IS SERIOUSLY EASY:
1) Water **EVAPORATES** off the sea.
2) Water **TRANSPIRES** from plants.
3) It turns to **CLOUDS** and falls as **RAIN**.
4) Then it **RUNS BACK TO THE SEA**.

FOUR EXTRA DETAILS:

(which are only very slightly harder to remember than the diagram)

1) The <u>SUN</u> causes the <u>evaporation</u> of water from the sea.

2) <u>Clouds form</u> because: when <u>air rises</u>, it <u>cools</u>, and the <u>water condenses</u> out.

3) When the condensed droplets get <u>too big</u> they <u>fall as rain</u>.

4) Some water is taken up by <u>roots</u> and <u>evaporates from trees</u> without ever reaching the <u>sea</u>.

There's not much to it but some long words

The only tricky bit here is remembering the fancy words, like "erosion" and "chemical weathering", and exactly what they are. Of course, you know "erosion" isn't quite the same as "weathering".

Plate Boundaries

At the boundaries between tectonic plates there's usually trouble like volcanoes or earthquakes.

Oceanic and Continental Plates Colliding: The Andes

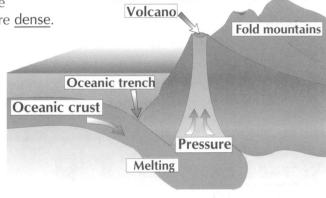

1) The oceanic plate is always forced underneath the continental plate, because oceanic plates are more dense.

2) This is called a subduction zone.

3) As the oceanic plate is pushed down, parts of it melt. The magma is less dense than the surrounding rock, so it rises.

4) Some molten rock finds its way to the surface and volcanoes form, but some cools slowly underground.

5) There are also earthquakes as the two plates slowly grind past each other.

6) A deep trench forms on the ocean floor where the oceanic plate is being forced down.

7) The continental crust crumples and folds forming mountains at the coast.

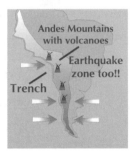

8) The classic example of all this is the west coast of South America where the Andes mountains are. That region has all the features:

VOLCANOES, EARTHQUAKES, an OCEANIC TRENCH and MOUNTAINS.

Two Continental Plates Collide: The Himalayas

1) The two continental plates meet head on, neither one being subducted.

2) Any sediment layers lying between the two continent masses get squeezed between them.

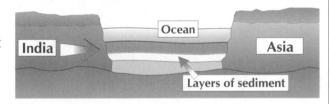

3) These sediment layers inevitably start crumpling and folding and soon form into big mountains.

4) The Himalayan mountains are the classic case of this.

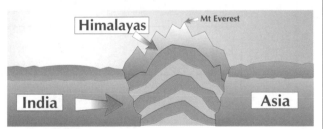

5) India actually broke away from the side of Africa and piled into the bottom of Asia, and is still doing so, pushing the Himalayas up as it goes.

6) Mount Everest is there and is getting higher by a few cm every year as India continues to push up into the continent of Asia.

Don't make a mountain out of an easy page of revision

Learn these diagrams — they summarise all the information in the text. They may well ask you for examples in the Exam, so learn the two different situations that the Andes and the Himalayas represent.

SECTION FOUR — AIR AND ROCK

Warm-Up and Worked Exam Questions

The warm-up questions run quickly over the basic facts you'll need in the Exam. The exam questions come later — but unless you've learnt the facts first you'll find the Exams tougher than old boots.

Warm-up Questions

1) Describe physical weathering.
2) Name two other types of weathering.
3) What does erosion mean?
4) Write down one of the geological features often seen at a plate boundary.
5) Describe how the Andes Mountains were formed.

Worked Exam Questions

There's a knack to be learnt in using the facts you've stored away in your brain box in the right way to get marks in the Exam. These worked examples will really help you see how...

1   Describe what you would find where the following movements of tectonic plates were occurring. For each one, give an example of a place where this is happening now.

   a)   an oceanic and a continental plate colliding

     Volcanoes, mountains, oceanic trench

     EXAMPLE *The Andes*

(2 marks)

   b)   two continental plates colliding

     Large mountains

     EXAMPLE *Himalayas*

(2 marks)

2   Rocks may be exposed to various forms of weathering and erosion. The following table describes some of them. Fill in the missing information in the table.

Description	Weathering type or Erosion?
Glaciers carving out a valley	Erosion
i) *Plants growing in a crack in a rock*	Biological weathering
River carrying away rocks from its bank	ii) *Erosion*
Marble headstone being eroded	iii) *Chemical weathering*

(3 marks)

Exam Questions

1   The diagram below shows the movement of two pieces of the Earth's crust.

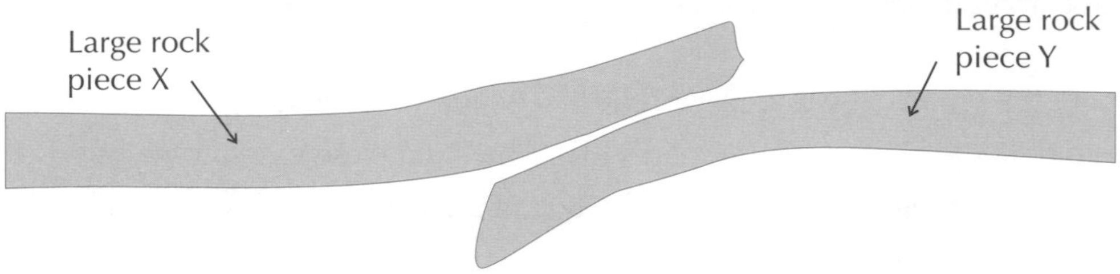

Large rock piece X

Large rock piece Y

a)   What is the name for these large pieces of rock?

....

(1 mark)

b)   i)   Rock pieces X and Y are moving towards one another. Suggest a reason why X is pushed over and above Y.

....

....

(1 mark)

ii)   Describe what might happen to rock piece Y as it is slowly pushed beneath X.

....

....

(1 mark)

2   In mountainous regions like the Cairngorms in Scotland, rocks are constantly being weathered.

a)   Name two types of weathering, and describe how they break up rocks into smaller fragments.

1. ....

....

(2 marks)

2. ....

....

(2 marks)

b)   The water cycle plays a very important part in processes of erosion.
    What is the water cycle's source of energy?

....

(1 mark)

Revision Summary

Well let's face it, this section on Air and Rock is definitely the easy interlude in the Chemistry syllabus. In the Olden Days (the 1970s) this stuff all used to be called Geography, which as you know, is a much easier subject than Chemistry. However, easy or not, there's still quite a lot of stuff to learn. Try these and see how much you know:

1) How old is the Earth? What was it like for the first billion years or so?

2) What gases did the early atmosphere consist of? Where did these gases come from?

3) What was the main thing which caused phase two of the atmosphere's evolution?

4) Which gases became much less common and which one increased?

5) Which gas allowed phase three to take place? Which gas is almost completely gone?

6) What are the percentages of gases in today's atmosphere?

7) Describe a simple experiment to find the percentage of oxygen in air.

8) Explain the three ways in which the oceans contain carbon, and two ways they contain salt.

9) Describe three man-made atmospheric problems.

10) Which gases cause acid rain? Where do they come from?
    What are the three adverse effects of acid rain?

11) Which gas causes the Greenhouse Effect? Explain how the Greenhouse Effect works.

12) Which type of gas is damaging the ozone layer? What are the harmful effects of this?

13) What is the carbon cycle? How much of that diagram on page 84 can you draw from memory?

14) What are the three types of rock? Draw a full diagram of the rock cycle.

15) Explain how the three types of rock change from one to another. How long does this take?

16) Draw diagrams to show how sedimentary rocks form.

17) What are found mainly in sedimentary rocks and only rarely in other types?

18) List the four main sedimentary rocks, give a description of each, and a use for two of them.

19) Draw a diagram to show how metamorphic rocks are formed. What does the name mean?

20) What are the three main metamorphic rocks?
    Describe their appearance and give a use for two of them.

21) How are igneous rocks formed? What are the two types? Give an example of each.

22) What is the difference in the way that they formed and in their structure and appearance?

23) What are the three types of weathering? Explain the details for each type, with diagrams.

24) What exactly is "erosion"? What is "transport"?

25) Describe how these processes created the Grand Canyon and other river valleys?

26) Draw a diagram of the water cycle. Explain how the whole process works.

27) What happens when an oceanic plate collides with a continental plate? Draw a diagram.

28) What four features does this produce? Which part of the world is the classic case of this?

29) What happens when two continental plates collide? Draw diagrams.

30) What features does this produce? Which part of the world is the classic case of this?

A Brief History of The Periodic Table

The early chemists were keen to try and find <u>patterns</u> in the elements.
The <u>more</u> elements that were identified, the <u>easier</u> it became to find patterns of course.

In the **Early 1800s** They Could Only go on **Atomic Mass**

They had <u>two</u> obvious ways to categorise elements:

1) Their <u>physical</u> and <u>chemical properties</u>	2) Their <u>Relative Atomic Mass</u>

1) Remember, they had <u>no idea</u> of <u>atomic structure</u> or of protons or electrons, so there was <u>no such thing</u> as <u>atomic number</u> to them. (It was only in the 20th Century after protons and electrons were discovered, that it was realised the elements should be arranged in order of <u>atomic number</u>.)

2) But <u>back then</u>, the only thing they could measure was <u>Relative Atomic Mass</u> and the only obvious way to arrange the known elements was <u>in order of atomic mass</u>.

3) When this was done a <u>periodic pattern</u> was noticed in the <u>properties</u> of the elements.

Newlands' Octaves Were The First Good Effort

A scientist called <u>Newlands</u> had the first good try at arranging the elements in <u>1863</u>. He noticed that every <u>eighth</u> element had similar properties and so he listed some of the known elements in rows of seven:

Li	Be	B	C	N	O	F
Na	Mg	Al	Si	P	S	Cl

These sets were called <u>Newlands' Octaves</u> but unfortunately the pattern <u>broke down</u> on the <u>third row</u> with many <u>transition metals</u> like Fe, Cu and Zn messing it up completely. It was because he left <u>no gaps</u> that his work was <u>ignored</u>. But he was getting <u>pretty close</u>, as you can see.

Dmitri Mendeleyev Left Gaps and **Predicted New Elements**

1) In <u>1869</u>, <u>Dmitri Mendeleyev</u> in Russia, armed with about 50 known elements, arranged them into his Table of Elements with various <u>gaps</u>, as shown.

2) Mendeleyev ordered the elements in order of <u>atomic mass</u> (like Newlands did).

3) But Mendeleyev found he had to leave <u>gaps</u> in order to keep elements with <u>similar properties</u> in the same <u>vertical groups</u> — and he was prepared to leave some <u>very big gaps</u> in the first two rows before the transition metals come in on the <u>fourth</u> row.

The <u>gaps</u> were the really clever bit because they <u>predicted</u> the properties of so far <u>undiscovered elements</u>.

When they were found and they <u>fitted the pattern</u> it was pretty smashing news for old Dmitri.

Mendeleyev's Table of the Elements

H																	
Li	Be												B	C	N	O	F
Na	Mg												Al	Si	P	S	Cl
K	Ca	*	Ti	V	Cr	Mn	Fe	Co	Ni	Cu	Zn	*	*	As	Se	Br
Rb	Sr	Y	Zr	Nb	Mo	*	Ru	Rh	Pd	Ag	Cd	In	Sn	Sb	Te	I
Cs	Ba	*	*	Ta	W	*	Os	Ir	Pt	Au	Hg	Tl	Pb	Bi

The History of Science

They're quite into having bits of History in Science now. They like to think you'll gain an appreciation of the role of science in the overall progress of human society. Whatever. Just learn it.

The Periodic Table

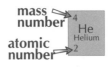
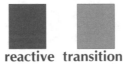

Group O

							1 H Hydrogen 1										4 He Helium 2

Group I   Group II

Group III   Group IV   Group V   Group VI   Group VII

2 7 Li Lithium 3 | 9 Be Beryllium 4 | | | | | | | | | | | 11 B Boron 5 | 12 C Carbon 6 | 14 N Nitrogen 7 | 16 O Oxygen 8 | 19 F Fluorine 9 | 20 Ne Neon 10

3 23 Na Sodium 11 | 24 Mg Magnesium 12 | | | | | | | | | | | 27 Al Aluminium 13 | 28 Si Silicon 14 | 31 P Phosphorus 15 | 32 S Sulphur 16 | 35.5 Cl Chlorine 17 | 40 Ar Argon 18

4 39 K Potassium 19 | 40 Ca Calcium 20 | 45 Sc Scandium 21 | 48 Ti Titanium 22 | 51 V Vanadium 23 | 52 Cr Chromium 24 | 55 Mn Manganese 25 | 56 Fe Iron 26 | 59 Co Cobalt 27 | 59 Ni Nickel 28 | 64 Cu Copper 29 | 65 Zn Zinc 30 | 70 Ga Gallium 31 | 73 Ge Germanium 32 | 75 As Arsenic 33 | 79 Se Selenium 34 | 80 Br Bromine 35 | 84 Kr Krypton 36

5 86 Rb Rubidium 37 | 88 Sr Strontium 38 | 89 Y Yttrium 39 | 91 Zr Zirconium 40 | 93 Nb Niobium 41 | 96 Mo Molybdenum 42 | 99 Tc Technetium 43 | 101 Ru Ruthenium 44 | 103 Rh Rhodium 45 | 106 Pd Palladium 46 | 108 Ag Silver 47 | 112 Cd Cadmium 48 | 115 In Indium 49 | 119 Sn Tin 50 | 122 Sb Antimony 51 | 128 Te Tellurium 52 | 127 I Iodine 53 | 131 Xe Xenon 54

6 133 Cs Caesium 55 | 137 Ba Barium 56 | 57-71 Lanthanides | 179 Hf Hafnium 72 | 181 Ta Tantalum 73 | 184 W Tungsten 74 | 186 Re Rhenium 75 | 190 Os Osmium 76 | 192 Ir Iridium 77 | 195 Pt Platinum 78 | 197 Au Gold 79 | 201 Hg Mercury 80 | 204 Tl Thallium 81 | 207 Pb Lead 82 | 209 Bi Bismuth 83 | 210 Po Polonium 84 | 210 At Astatine 85 | 222 Rn Radon 86

7 223 Fr Francium 87 | 226 Ra Radium 88 | 89-103 Actinides

mass number → 4 He Helium

atomic number → 2

| reactive metals | transition elements | poor metals | non metals | noble gases | separates metals from non-metals |

The *Periodic Table*

1) There are 100ish elements, which all materials are made of. More are still being discovered.

2) The <u>modern</u> Periodic Table shows the elements in order of <u>atomic number</u>.

3) The Periodic Table is laid out so that elements with <u>similar properties</u> form <u>columns</u>.

4) These <u>vertical columns</u> are called <u>Groups</u> and Roman Numerals are often used for them.

5) For example the <u>Group II</u> elements are Be, Mg, Ca, Sr, Ba and Ra.
They're all <u>metals</u> which form 2+ ions and they have many other <u>similar properties</u>.

6) The <u>rows</u> are called <u>periods</u>. Each <u>new period</u> represents <u>another full shell</u> of electrons.

So, all the elements in a Group have the
<u>same number of electrons in the outer shell,</u>
<u>but different numbers of shells.</u>
For example, these Group I atoms:

One electron in
the outer shell

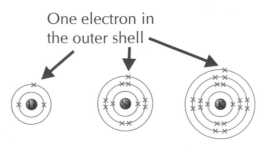

The periodic table is all you need...

Not knowing the periodic table is a bit like not knowing the Highway Code. You can try to do chemistry without it, but it's likely to all end in disaster, heaps of twisted metal, and expensive garage bills. Learn the rules, know the trends and drive safely.

The Periodic Table

The Elements of a **Group** Have the **Same Outer Electrons**

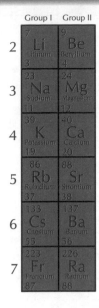

1) The elements in each <u>Group</u> all have the same number of <u>electrons</u> in their <u>outer shell</u>.

2) That's why they have <u>similar properties</u>. And that's why we arrange them in this way.

3) You absolutely must get that into your head if you want to <u>understand</u> any Chemistry.

The properties of the elements are decided *entirely* by how many electrons they have. Atomic number is therefore very significant because it is equal to how many electrons each atom has. But it's the number of electrons in the <u>outer shell</u> which is the really important thing.

Electron Shells are really important

The fact that electrons form shells around atoms is the reason for the whole of chemistry. If they just whizzed round the nucleus any old how and didn't care about shells or any of that stuff there'd be no chemical reactions. No nothing in fact — because nothing would happen.

Without shells there'd be no atoms wanting to gain, lose or share electrons to form full shell arrangements. So they wouldn't be interested in forming ions or covalent bonds. There'd be no molecules and no reactions. Nothing would bother and nothing would happen. The atoms would just sit around, all day long doing nothing.

But amazingly, they *do* form shells (if they didn't, we wouldn't even be here to wonder about it), and the electron arrangement of each atom determines the whole of its chemical behaviour.

Phew. I mean electron arrangements explain practically the whole Universe. They're just great.

Electron Shells — where would we be without them

...we'd be up a really smelly river without any sort of rowing implement. So make sure you learn the whole periodic table including every name, symbol and number. No, only kidding. Just <u>learn</u> the numbered points on the last two pages and <u>scribble</u> them down, <u>mini-essay style</u>.

Electron Arrangements

This diagram shows the <u>electron arrangements</u> of the first <u>twenty</u> elements.
Learn it all right now.

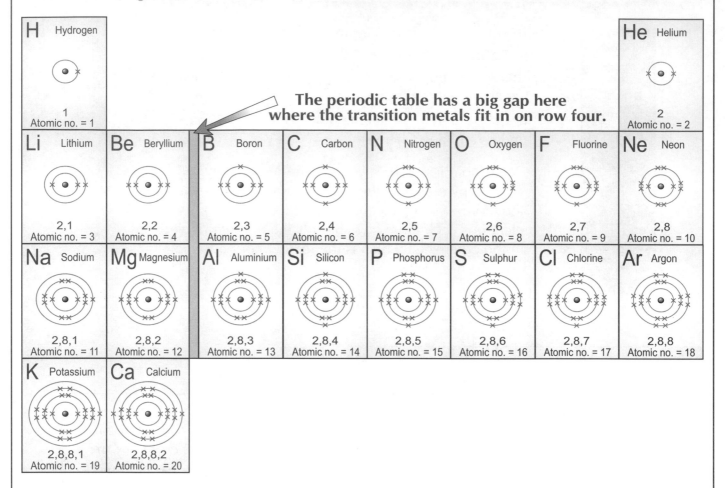

**The periodic table has a big gap here
where the transition metals fit in on row four.**

Reactivity Changes down the Groups due to Shielding

1) As atoms get <u>bigger</u>, they have <u>more full shells</u> of electrons.

2) As you go down any Group, each <u>new row</u> has <u>one more</u> full shell.

3) The number of <u>outer</u> electrons is the <u>same</u> for each element in a Group.

4) However the outer shell of electrons is <u>increasingly far</u> from the nucleus.

5) You have to learn to say that the inner shells provide "<u>SHIELDING</u>".

6) This means that the <u>outer shell electrons</u> get <u>shielded</u> from the <u>attraction</u>
   of the <u>+ve nucleus</u>.

Shielding must be learnt

Shielding is a thoroughly useful concept which explains a lot about why different elements do what
they do. Make sure you can talk confidently about the concept of shielding. Of course, the best way to
do this is to read the points on this page, cover them up and then scribble them down. Well, go on.

Electron Arrangements

Metals get more reactive the larger they get

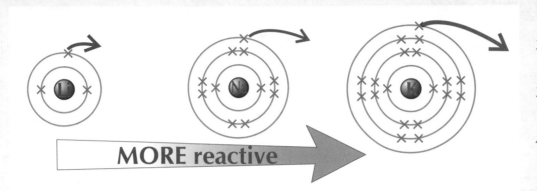

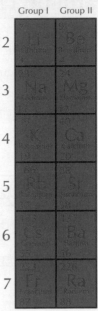

As metal atoms get bigger, the outer electron is more easily lost. This is due to shielding. This makes METALS MORE REACTIVE as you go DOWN Group I and Group II.

Non-Metals get less reactive the larger they get

As non-metal atoms get bigger, the extra electrons are harder to gain. This makes NON-METALS LESS REACTIVE as you go DOWN Group VI and Group VII

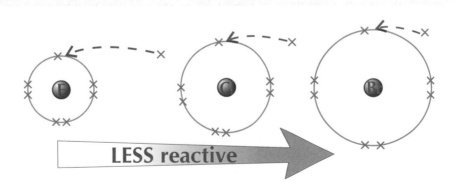

Learn about Electron Shielding — and keep up with the trends...

It's really important that you know how electron arrangement relates to an element's position in the periodic table. Obviously you don't learn every atom separately — you learn the pattern. Also learn about the trends in reactivity. Then cover the page and see what you know — by scribbling.

Warm-Up and Worked Exam Questions

The Periodic Table is really all about electron arrangements. There'll almost certainly be a question about the electron arrangement of an element, and how that relates to chemical properties.

Warm-up Questions

1) True or false? — elements in the same row of the Periodic Table have similar chemical properties.
2) How many electrons are there in the innermost shell of carbon, nitrogen and calcium?
3) How many electrons are there in the outer shell of a Group 1 element?
4) Which of the following elements has very different chemical properties to the other three: fluorine; iodine; bromine; aluminium?

Worked Exam Question

Here's a worked question about electron arrangements in the Periodic Table.

1   This diagram shows the electron arrangement of an atom.

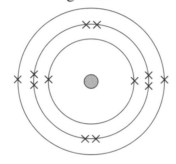

a)   What element does this diagram show?

.... *Magnesium* .... *Count the electrons, then find the element* ....
with the same atomic number.
(1 mark)

b)   Use the diagram to explain why this element is in Group II of the Periodic Table.

.... *Because it has two electrons in its outer shell.* ....

....
(1 mark)

c)   This diagram shows beryllium.

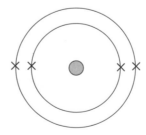

Explain why beryllium is less reactive than the element shown in the first diagram.
Beryllium has fewer electron shells than magnesium. Its outer
electrons are less shielded from the attraction of the positive
nucleus. Beryllium therefore gives up its outer electrons less readily.
(3 marks)

Exam Questions

1   Before the modern Periodic Table was designed, scientists tried various ways to order the elements. In 1863, John Newlands designed a table that showed some of the known elements arranged in groups of seven.

Li	Be	B	C	N	O	F
Na	Mg	Al	Si	P	S	Cl

a)   Which element in the table has similar chemical properties to oxygen?

....
(1 mark)

b)   Did Newlands have any idea of atomic structure or electron arrangement when he made his table of elements?

....
(1 mark)

2   Look at these diagrams of electron arrangements in atoms of three different elements.

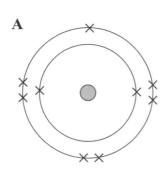 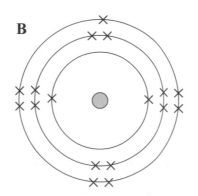 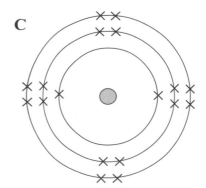

a)   Which two elements form ionic salts with Group I and Group II metals?

....
(1 mark)

b)   Which element is unreactive?

....
(1 mark)

c)   Which element is the most reactive?

....
(1 mark)

Group 0 — The Noble Gases

Group O

Group VI	Group VII	Group O
		4 He Helium 2
O	F	20 Ne Neon 10
S	Cl	40 Ar Argon 18
Se	Br	84 Kr Krypton 36
Te	I	131 Xe Xenon 54
Po	At	222 Rn Radon 86

As you go **down** the Group:

1) The density increases because the atomic mass increases.

2) The boiling point increases
Helium boils at –269°C (that's very cold)
Xenon boils at –108°C (that's still very cold)

They all have full outer shells —

That's why they're so inert.

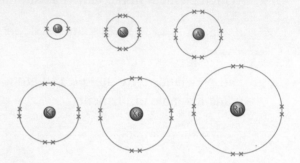

Helium, Neon and Argon are Noble gases

There's also <u>Krypton</u>, <u>Xenon</u> and <u>Radon</u>, which may get asked for.
They're also sometimes called the <u>Inert</u> gases. Inert means "doesn't react".

They're all colourless, monatomic gases

<u>Most</u> gases are made up of <u>molecules</u>, but these <u>only exist</u> as <u>individual atoms</u>, because they <u>won't form bonds</u> with anything.

The Noble gases don't react at all

Helium, Neon and Argon don't form <u>any kind of chemical bonds</u> with anything.
They <u>always</u> exist as separate atoms. They won't even join up in pairs.

Helium is used in Airships and Party Balloons

Helium is ideal: it has very <u>low density</u> and <u>won't set on fire</u> (like hydrogen does).

Neon is used in electrical discharge tubes

When a current is passed through neon it gives out a bright light.

Argon is used in filament lamps (light bulbs)

It provides an <u>inert atmosphere</u> which stops the very hot filament from <u>burning away</u>.

All three are used in lasers too

There's the famous little red <u>Helium-Neon</u> laser and the more powerful <u>Argon laser</u>.
(You get Krypton lasers too.)

Noble, but lazy...

Well they don't react so there's obviously not much to learn about these. Nevertheless, there's likely to be a question or two on them so <u>make sure you learn everything on this page</u>.

Group I — The Alkali Metals

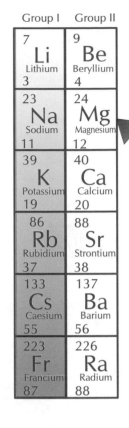

Group I	Group II
7 **Li** Lithium 3	9 **Be** Beryllium 4
23 **Na** Sodium 11	24 **Mg** Magnesium 12
39 **K** Potassium 19	40 **Ca** Calcium 20
86 **Rb** Rubidium 37	88 **Sr** Strontium 38
133 **Cs** Caesium 55	137 **Ba** Barium 56
223 **Fr** Francium 87	226 **Ra** Radium 88

These *Group II* metals are quite similar to Group I, except that they have two electrons in the outer shell and form 2+ ions. They are less reactive.

Learn These Trends:

As you go DOWN Group I, the Alkali Metals:

1) become Bigger Atoms
...because there's one extra full shell of electrons for each row you go down.

2) become More Reactive
...because the outer electron is more easily lost, because it's further from the nucleus.

3) have a Higher Density
...because the atoms have more mass.

4) become Even Softer to Cut

5) have a Lower Melting Point

6) have a Lower Boiling Point

1) The Alkali Metals are very **Reactive**

They have to be <u>stored in oil</u> because they would react with air or water. They must be handled with <u>forceps</u> (otherwise they burn the skin).

2) They are: **Lithium**, **Sodium**, **Potassium** and a couple more

Know those three names really well. They may also mention Rubidium and Caesium.

3) The Alkali Metals all have **ONE outer electron**

This makes them very <u>reactive</u> and gives them all similar properties. They're all very close to having full electron shells and so give up their one outer electron very readily.

Learn about Alkali Metals — or get your fingers burnt...

Now we're getting into the seriously dreary facts section. This takes a bit of learning, this stuff does, especially those trends in behaviour as you go down the Group. <u>Enjoy</u>.

Group I — The Alkali Metals

4) The Alkali Metals all form 1⁺ ions

They are <u>keen to lose</u> their one outer electron to form a <u>1⁺ ion</u>:

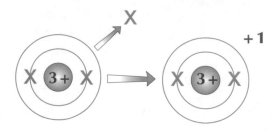

5) The Alkali metals always form White Ionic Compounds

They are so keen to lose the outer electron there's <u>no way</u>
they'd consider <u>sharing</u>, so covalent bonding is <u>out of the question</u>.
The ionic compounds are <u>white</u>, but dissolved in water they're <u>colourless</u>.

6) The Alkali metals are soft — they cut with a knife

Lithium is the hardest, but still easy to cut with a scalpel.
They're <u>shiny</u> when freshly cut, but <u>soon go dull</u> as they
react with the air.

7) The Alkali metals melt and boil easily (for metals)

Lithium melts at 180°C , Caesium at 29°C.
Lithium boils at 1330°C, Caesium at 670°C

8) The Alkali metals have low density (they float)

Lithium, Sodium and Potassium are all <u>less dense than water</u>.
The others "<u>float</u>" anyway, on the H_2 bubbles they produce when
they react with water.

Group I elements are not your average metals

Generally speaking, all the elements in Group I are pretty strange. They're all metallic, and shiny when
you cut them, but they're also all squidgy and soft. They react with water, so remember NOT to throw
them down the sink after your experiment. You'll see what I mean on the next page.

Reactions of the Alkali Metals

Reaction with Cold Water produces **Hydrogen Gas**

1) When <u>lithium</u>, <u>sodium</u> or <u>potassium</u> are put in <u>water</u>, they react very <u>vigorously</u>.

2) They <u>move</u> around the surface, <u>fizzing</u> furiously.

3) They produce <u>hydrogen</u>. Potassium gets hot enough to <u>ignite</u> it.
   A lighted splint will <u>indicate</u> hydrogen by producing the notorious
   "<u>squeaky pop</u>" as the H_2 ignites.

4) Sodium and potassium <u>melt</u> in the heat of the reaction.

5) They form a <u>hydroxide</u> in solution, i.e. <u>aqueous OH$^-$ ions</u>.

$$2Na_{(s)} + 2H_2O_{(l)} \rightarrow 2NaOH_{(aq)} + H_{2\,(g)}$$

$$2K_{(s)} + 2H_2O_{(l)} \rightarrow 2KOH_{(aq)} + H_{2\,(g)}$$

The solution becomes <u>alkaline</u>,
which changes the colour of
the pH indicator to <u>purple</u>.

Alkali Metal **Oxides** and **Hydroxides** are **Alkaline**

This means that they'll react with <u>acids</u> to form <u>neutral salts</u>, like this:

$$NaOH + HCl \rightarrow H_2O + NaCl \text{ (salt)}$$

$$Na_2O + 2HCl \rightarrow H_2O + 2NaCl \text{ (salt)}$$

All Alkali Metal **Compounds** look like **Salt** and **Dissolve** readily

1) All alkali metal compounds are <u>ionic</u>, so they form <u>crystals</u> which <u>dissolve</u> easily.

2) They're all very <u>stable</u> because the alkali metals are so <u>reactive</u>.

3) Because they always form <u>ionic</u> compounds with <u>giant ionic</u>
   <u>lattices</u> the compounds <u>all</u> look pretty much like the regular
   '<u>salt</u>' you put on your chips.

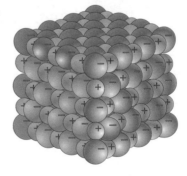

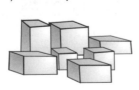

The Notorious Squeaky Pop...

This stuff's pretty tough isn't it. Think you can learn it, punk? Well let's just see you try...
If you keep covering the page and repeating bits back to yourself, or scribbling bits down, then little by
little <u>it does go in</u>. Little by little.

Warm-Up and Worked Exam Questions

Questions about the reactivity of alkali metals or inert gases are fairly easy once you know that full outer electron shells are stable. There are a few equations to get right, too, so don't be careless.

Warm-up Questions

1) What ion is formed by lithium?
2) Does helium form any compounds?
3) Why does neon exist as a monatomic gas while fluorine and oxygen are diatomic gases?
4) Is the boiling point of potassium low or high compared to the average metal?
5) Name a metal which floats on water.

Worked Exam Questions

Expect to get some kind of question about alkali metals in the Exam.

1 (a) Why must sodium be kept under oil?

The oil stops air reaching the sodium. Sodium would react with

oxygen in the air.   Make sure you say how the oil stops the
                         sodium reacting.

(2 marks)

(b) Sodium can be cut with a knife.

(i) When sodium is cut, the cut surface becomes dull and whitish. Explain why.

The cut surface reacts with oxygen in the air to become

sodium oxide, which is white.

(1 mark)

(ii) Which of the alkali metals is the hardest?

Lithium   They get softer down the group, so lithium must
            be the hardest one.   *(1 mark)*

(c) The elements in Group I get more reactive the further down the group you go. Explain this trend in terms of electron arrangement and ease of ion formation.

As you go down the group, the outer electron is more easily lost

to form an ion because it's further from the nucleus and it's

Distance from nucleus and   *shielded from the attraction of the +ve*
shielding are THE two
factors to mention.   *nucleus by the inner electrons.*

(3 marks)

2 Explain why the noble gases are also called the inert gases.

They're inert. They don't react because they have a stable full

outer electron shell.   Full outer shell = stable = no need to lose or gain
                           electrons by reacting.

(2 marks)

Exam Questions

1   Potassium reacts violently when it is dropped into a beaker of cold water.

  (a) Write down the two products of this reaction.

  ....

  (1 mark)

  (b) Write a balanced equation for the reaction.

  ....

  (1 mark)

  (c) Sodium reacts less violently than potassium when it is dropped into water.
       Explain why in terms of electron arrangement.

  ....

  ....

  ....

  (3 marks)

2   This question is about the noble gases.

  (a) The noble gases are also called the inert gases, because they do not react.

    (i) Explain why the noble gases do not form compounds.

    ....

    ....

    (1 mark)

    (ii) Oxygen gas is found as a diatomic molecule O_2. In what form does neon exist?

    ....

    (1 mark)

  (b) The noble gases are used in lighting.

    (i) Which noble gas is used in fluorescent tube lighting?

    ....

    (1 mark)

    (ii) Argon is used in filament light bulbs. Explain why.

    ....

    (1 mark)

  (c) The *Hindenburg* was a hydrogen-filled airship which exploded disastrously in 1935.
       After the disaster, helium was used to fill airships instead of hydrogen. Why was this?

  ....

  ....

  (2 marks)

Group VII — The Halogens

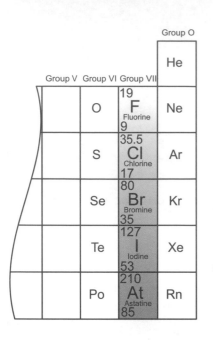

Learn These Trends:

As you go *DOWN* Group VII, the *HALOGENS*:

1) become Bigger Atoms
...because there's one extra full shell of electrons for each row you go down.

2) become Less Reactive
...because there's less inclination to gain the extra electron to fill the outer shell when it's further out from the nucleus.

3) become Darker in Colour

4) go from Gas to Solid
Fluorine and chlorine are gases, bromine is a liquid, and iodine is a solid.

5) have a Higher Melting Point

6) have a Higher Boiling Point

1) The Halogens are all non-metals with coloured vapours

<u>Fluorine</u> is a very reactive, poisonous <u>yellow gas</u>.

<u>Chlorine</u> is a fairly reactive, poisonous <u>dense green gas</u>.

<u>Bromine</u> is a dense, poisonous, <u>red-brown volatile liquid</u>.

<u>Iodine</u> is a <u>dark grey</u> crystalline <u>solid</u> or a <u>purple vapour</u>.

2) They all form molecules which are pairs of atoms:

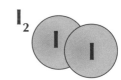

F_2 Cl_2 Br_2 I_2

3) The Halogens do both ionic and covalent bonding

The Halogens all form <u>ions</u> <u>with a 1⁻ charge</u>: F^- Cl^- Br^- I^- as in Na^+Cl^- or $Fe^{3+}Br^-_3$

They form <u>covalent bonds</u> with <u>themselves</u> and in various <u>molecular compounds</u> like these:

<u>Carbon tetrachloride:</u>
(CCl_4)

<u>Hydrogen chloride:</u>
(HCl)

4) The Halogens are poisonous — always use a fume cupboard

What more can I say? Use a fume cupboard, or else...

Halogens are just one electron short of a full shell
Well, I think Halogens are just slightly less grim than the Alkali metals. At least they change colour and go from gases to liquid to solid. <u>Learn all the little facts anyway</u>.

Reactions of the Halogens

1) The Halogens react with **metals** to form **salts**

They react with most metals including iron and aluminium, to form salts (or "metal halides").

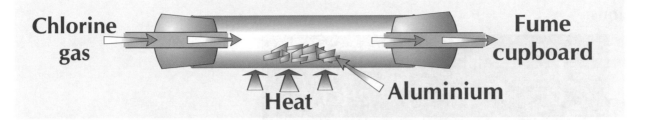

Chlorine gas → **Fume cupboard**

Heat **Aluminium**

Equations:

$$2Al_{(s)} + 3Cl_{2\ (g)} \rightarrow 2AlCl_{3\ (s)} \quad \text{(Aluminium chloride)}$$
$$2Fe_{(s)} + 3Br_{2\ (g)} \rightarrow 2FeBr_{3\ (s)} \quad \text{(Iron(III) bromide)}$$

2) More reactive Halogens will **displace** less reactive ones

Chlorine can displace bromine and iodine from a solution of bromide or iodide. Bromine will also displace iodine because of the trend in reactivity.

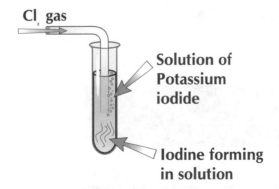

Cl$_2$ gas

Solution of Potassium iodide

Iodine forming in solution

Equations:

$$Cl_{2\ (g)} + 2KI_{(aq)} \rightarrow I_{2\ (aq)} + 2KCl_{(aq)}$$
$$Cl_{2\ (g)} + 2KBr_{(aq)} \rightarrow Br_{2\ (aq)} + 2KCl_{(aq)}$$

More Halogens...

I'm afraid it's time to learn more equations. Alas, there's no way round it, just get stuck in and commit them to memory. Now cover the page and try scribbling them down. Better safe than sorry.

Reactions of the Halogens

3) Halogens **react** with hydrogen

The halogens form <u>covalent compounds</u> with <u>hydrogen</u>.
Fluorine and chlorine can explode as they react with hydrogen.
Bromine and iodine only react slowly.

Equations:

$$F_{2(g)} + H_{2(g)} \rightarrow 2HF_{(g)}$$
$$Cl_{2(g)} + H_{2(g)} \rightarrow 2HCl_{(g)}$$

4) Hydrogen Chloride **gas** dissolves to form **HCl acid**

1) <u>Hydrogen chloride</u> is a <u>diatomic molecule</u> (a two-atom molecule), held together by a <u>covalent</u> bond.

2) It has a <u>simple molecular</u> structure.

3) It is a <u>dense</u>, <u>colourless gas</u> with a <u>choking</u> smell.

4) Gaseous hydrogen chloride is important in the manufacture of <u>polymers</u>.

5) It <u>dissolves</u> in water, which is very <u>unusual</u> for a <u>covalent</u> substance, to form the <u>well-known</u> strong acid, <u>hydrochloric acid</u>.

6) The <u>proper method</u> for dissolving hydrogen chloride in water is to use an <u>inverted funnel</u> as shown:

7) HCl gas <u>reacts with water</u> to produce <u>H^+ ions</u>, which is what makes it <u>acidic</u>:

Hydrogen Chloride   Covalent bond

Hydrogen Chloride

$$HCl_{(g)} \xrightarrow{\text{water}} H^+_{(aq)} + Cl^-_{(aq)}$$

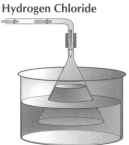

Hydrogen Bromide and Hydrogen Iodide do the same

Just like hydrogen chloride, these two <u>gases</u> will also <u>dissolve easily</u> to form <u>strong acids</u>:

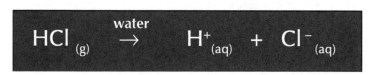

$$HBr_{(g)} \rightarrow H^+_{(aq)} + Br^-_{(aq)} \qquad HI_{(g)} \rightarrow H^+_{(aq)} + I^-_{(aq)}$$

Salts and Acids — what an unsavoury combination...

More exciting reactions to delight and entertain you through the shove and shuffle of your otherwise dreary teenage years. Learn the equations and then bask in the glow of your new-found wisdom. Pity those who are deprived of Chemistry.

Industrial Salt

Salt is taken from *the sea* — and from *underneath Cheshire*

1) In <u>hot countries</u> they just pour <u>sea water</u> into big flat open tanks and let the <u>sun</u> evaporate the water to leave salt. This is no good in cold countries because there isn't enough sunshine.

2) In <u>Britain</u> (a cold country — as if you need reminding), salt is extracted from <u>underground deposits</u> left <u>millions of years</u> ago when <u>ancient seas</u> evaporated. There are massive deposits of this <u>ROCK SALT</u> in <u>Cheshire</u>. It's taken from <u>underground mines</u>. Rock salt is a mixture of mainly <u>sand</u> and <u>salt</u>. It can be used in its <u>raw state</u> on roads, or the salt can be filtered out for more <u>refined uses</u>, as detailed below.

1) *Rock salt* is used for *de-icing roads*

1) The <u>salt</u> in the mixture <u>melts ice</u> by <u>lowering the freezing point</u> of water to around –5˚C.

2) The <u>sand</u> and <u>grit</u> in it gives useful <u>grip</u> on ice which hasn't melted.

2) *Salt* (sodium chloride) is used in the *food industry,* somewhat

<u>Salt</u> is added to most <u>processed foods</u> to enhance the <u>flavour</u>. It's now reckoned to be <u>unhealthy</u> to eat too much salt.

> *I'm just waiting for the great day of reckoning when finally every single food has been declared either generally unhealthy or else downright dangerous. Perhaps we should all lay bets on what'll be the last food still considered safe to eat.*
> *My money's on Dried Locusts.*

3) *Salt* is used for making *chemicals*

It all starts with the electrolysis of brine...

Salt — more useful than you ever imagined

There's only a few really important points on this page — just where salt comes from and a bit about some of its uses. The next page has the third and most exciting use of salt. Well I guess it's only really exciting if you want to know where all the lovely chemicals come from — and who wouldn't.

Industrial Salt

<u>Salt</u> is important for the <u>chemicals industries</u>, which are mostly based around <u>Cheshire</u> and <u>Merseyside</u> because of all the <u>rock salt</u> there. The first thing they do is <u>electrolyse</u> the salt like this:

Electrolysis of Salt gives Hydrogen, Chlorine and NaOH

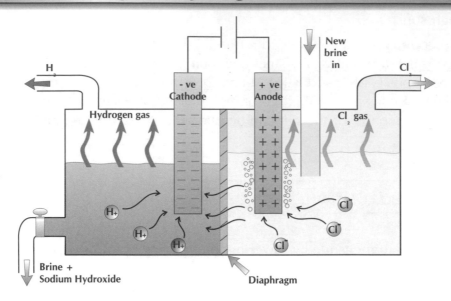

<u>Salt</u> dissolved in <u>water</u> is called <u>BRINE</u>. When <u>concentrated brine</u> is <u>electrolysed</u> there are <u>three useful products</u>:

a) <u>Hydrogen gas</u> is given off at the cathode.

b) <u>Chlorine gas</u> is given off at the anode.

c) <u>Sodium hydroxide</u> is left in solution.

These are collected, and then <u>used</u> in all sorts of <u>industries</u> to make various <u>products</u>.

Useful Products from the Electrolysis of Brine

1) Chlorine

1) Used in <u>bleach</u>, for <u>sterilising water</u> and for making <u>HCl</u> and <u>insecticides</u>.
2) Used to make <u>CFCs</u> for <u>fridges</u>, <u>aerosols</u> and <u>plastics</u>, but they're used a lot less now we know they damage the <u>ozone layer</u>.

2) Hydrogen

1) Used in the <u>Haber Process</u> to make <u>ammonia</u>.
2) Used to change <u>oils</u> into <u>fats</u> for making <u>margarine</u> ("hydrogenated vegetable oil").

3) Sodium hydroxide

Sodium Hydroxide is a <u>very strong alkali</u> and is <u>used widely</u> in the <u>chemical industry</u> to make, e.g.

1) soap   2) ceramics   3) organic chemicals   4) paper pulp   5) oven cleaner.

Learn the many uses of salty water

Brine is not just for going in outdoor swimming pools and storing tuna fish in. Now you know where all those wonderful chemicals you know and love come from. They were in brine all the time.

Uses of Halogens and Salt Products

Some *Uses* of *Halogens* you Really Should Know

Fluorine *(or rather **fluoride**) reduces **dental decay***

Fluorides can be added to drinking water and toothpastes to help prevent tooth decay.

Chlorine *is used in **bleach** and for **sterilising water***

1) Chlorine dissolved in sodium hydroxide solution is called bleach.
2) Chlorine compounds are also used to kill germs in swimming pools and drinking water.
3) It's also used to make insecticides and in the manufacture of HCl.

Iodine *is used as an **antiseptic**...*

...but it stings like nobody's business and stains the skin brown.

Silver halides *are used on black and white **photographic film***

1) Silver is very unreactive. It does form halides but they're very easily split up.
2) In fact, ordinary visible light has enough energy to do so.
3) Photographic film is coated with colourless silver bromide.
4) When light hits parts of it, the silver bromide splits up into silver and bromine:

$$2AgBr \rightarrow Br_2 + 2Ag \text{ (silver metal)}$$

5) The silver metal appears black. The brighter the light, the darker it goes.
6) This produces a black and white negative, like an X-ray picture for example.

You'll need to know the uses of the Halogens

You know how Examiners like asking you for examples? Well if you know, then why haven't you learnt this page off by heart? Come on, some of it's quite exciting — look at that lovely bit about iodine...

Warm-Up and Worked Exam Questions

You could read through this page in a few minutes but there's no point unless you check over any bits you don't know and make sure you understand everything. It's not quick but it's the only way.

Warm-up Questions

1) What colour is fluorine gas?
2) In what physical state is bromine at room temperature and pressure?
3) Which common halide is added to food to enhance flavour?
4) Why is chlorine added to swimming pools?
5) Why is fluorine added to tap water?

Worked Exam Question

Take your time to go through this example and make sure you understand it all. If any of the facts are baffling you it's not too late to take another peek over the section.

1   a)   Describe the physical appearance of the following elements at r.t.p. *(room temperature and pressure)*

   i)   Chlorine

   A greenish gas

   (2 marks)

   ii)   Iodine

   A grey solid producing purple vapour

   (3 marks)

   ii)   Bromine

   A reddish brown liquid

   (2 marks)

   b)   Write down a balanced equation for the reaction between aluminium and chlorine gas.

   $2Al + 3Cl_2 \rightarrow 2AlCl_3$ *Remember to make it balanced with the same number of atoms on each side for each element.*   *(2 marks)*

   c)   Chlorine can form ionic or covalent bonds.

   i)   Give an example of a covalently bonded molecule containing chlorine.

   Chlorine gas, Cl_2   *Or you could have carbon tetrachloride, CCl_4*

   (1 mark)

   ii)   Give an example of an ionically bonded compound containing chlorine.

   Sodium chloride, NaCl   *Or any chloride you like...*

   (1 mark)

Exam Questions

1   This question is about the extraction and uses of sodium chloride and other halides.

   a)   Rock salt is used for de-icing roads. Explain why.

   ....

   ....

   ....

   (2 marks)

   b)   Describe how salt is extracted from sea water in hot countries.

   ....

   ....

   (1 mark)

   c)   Silver halides are used in photographic film.

   i)   Write a balanced equation for what happens when silver bromide is exposed to light.

   ....

   (1 mark)

   ii)   What colour does the photographic film appear where the light has hit it?

   ....

   (1 mark)

2   This question is about making hydrochloric acid.
   Hydrogen gas and chlorine gas react together to make hydrogen chloride.

$$Cl_{2(g)} + H_{2(g)} \rightarrow 2HCl_{(g)}$$

   a)   How is hydrogen chloride gas bonded?

   ....

   (1 mark)

   b)   Hydrogen chloride gas dissolves very readily in water.

   i)   Write an equation to show this.

   ....

   (2 marks)

   ii)   Which product of this reaction makes the solution acidic?

   ....

   (1 mark)

Exam Questions

3   This question is about Group VII of the Periodic Table, the halogens.

(a)   Explain why reactivity decreases as you go down Group VII.

....

....

....

(2 marks)

(b)   Halogens form ionic compounds with alkali metals.

(i)   Chlorine gas will react with potassium iodide to form iodine and potassium chloride. Write a balanced equation for this reaction.

....

(2 marks)

(ii)   Will iodine react with potassium chloride to form chlorine and potassium iodide? Explain your answer.

....

....

(2 marks)

4   Concentrated brine is electrolysed to give three useful products.

(a)   Brine is a solution of which halide salt?

....

(1 mark)

(b)   Write down the name of each of the three products of the industrial electrolysis of concentrated brine, and give an industrial use for each one.

(i)   Name: ....

Use:

....

(2 marks)

(ii)   Name: ....

Use:

....

(2 marks)

(iii)   Name: ....

Use:

....

(2 marks)

Acids and Alkalis

The *pH Scale* and *Universal Indicator*

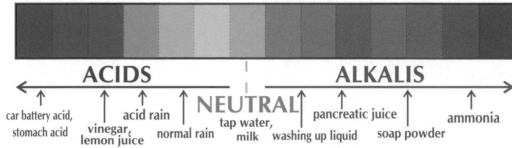

An *Indicator* is just a *Dye* that *changes colour*

The dye <u>changes colour</u> depending on whether it's <u>in an acid</u> or <u>in an alkali</u>.
<u>Universal indicator</u> is a very useful <u>combination of dyes</u> which give the colours shown above.

The *pH scale* goes from *1 to 14*

1) The <u>strongest acid</u> has <u>pH 1</u>. The <u>strongest alkali</u> has <u>pH 14</u>.
2) If something is <u>neutral</u> it has <u>pH 7</u> (e.g. pure water).
3) Anything <u>less</u> than 7 is <u>acid</u>. Anything <u>more</u> than 7 is <u>alkaline</u>.
   (An alkali can also be called a base.)

Acids have H⁺ *ions*   *Alkalis* have OH⁻ *ions*

The <u>strict definitions</u> of acids and alkalis are:

> <u>ACIDS</u> are substances which form <u>$H^+_{(aq)}$ ions</u> when added to *water*.
> <u>ALKALIS</u> are substances which form <u>$OH^-_{(aq)}$ ions</u> when added to *water*.

Neutralisation

This is the equation for <u>any neutralisation reaction</u>. Make sure you learn it:

$$Acid + alkali \rightarrow salt + water$$

Neutralisation can also be seen <u>in terms of ions</u> like this — so learn it too:

$$H^+_{(aq)} + OH^-_{(aq)} \rightarrow H_2O_{(l)}$$

Three *"Real life"* Examples of *Neutralisation*:

1) <u>Indigestion</u> is caused by too much <u>hydrochloric acid</u> in the <u>stomach</u>.
   <u>Indigestion tablets</u> contain <u>alkali</u> such as <u>magnesium oxide</u>, which <u>neutralise</u> the <u>excess HCl</u>.
2) <u>Fields</u> with <u>acidic soils</u> can be improved no end by adding <u>lime</u> (see page 51).
   The lime added to fields is <u>calcium hydroxide</u> $Ca(OH)_2$ which is of course an <u>alkali</u>.
3) <u>Lakes</u> affected by <u>acid rain</u> can also be <u>neutralised</u> by adding <u>lime</u>. This saves the fish.

Acids form H⁺ ions
Try and enjoy this page on acids and alkalis, because it gets harder from here on in. These are very basic facts and possibly quite interesting. <u>Cover the page and scribble them down</u>.

Acids Reacting with Metals

Acid + Metal → Salt + Hydrogen

That's written big because it's kind of worth remembering. Here's the <u>typical experiment</u>:

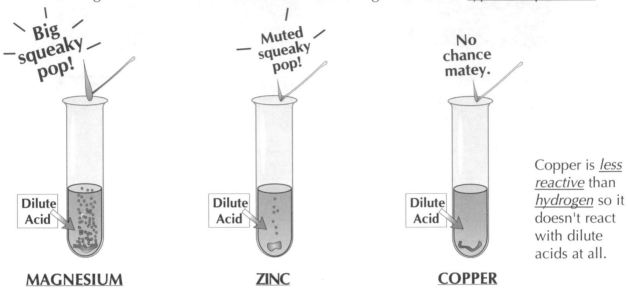

Big squeaky pop!   Muted squeaky pop!   No chance matey.

Dilute Acid   Dilute Acid   Dilute Acid

MAGNESIUM   **ZINC**   **COPPER**

Copper is *less reactive* than *hydrogen* so it doesn't react with dilute acids at all.

1) The <u>more reactive</u> the metal, the <u>faster</u> it will go.
2) <u>Copper</u> does <u>not</u> react with dilute acids <u>at all</u> — because it's <u>less reactive than hydrogen</u>.
3) The <u>speed of reaction</u> is indicated by the <u>rate</u> at which the <u>bubbles of hydrogen</u> are given off.
4) The <u>hydrogen</u> is confirmed by the <u>burning splint test</u> giving the notorious "<u>squeaky pop</u>".
5) The <u>type of salt</u> produced depends on which <u>metal</u> is used, and which <u>acid</u> is used:

Hydrochloric acid will always produce **chloride** salts:

$2HCl + Mg \rightarrow MgCl_2 + H_2$ (Magnesium chloride)
$2HCl + Zn \rightarrow ZnCl_2 + H_2$ (Zinc chloride)

Sulphuric acid will always produce **sulphate** salts:

$H_2SO_4 + Mg \rightarrow MgSO_4 + H_2$ (Magnesium sulphate)
$H_2SO_4 + Zn \rightarrow ZnSO_4 + H_2$ (Zinc sulphate)

Nitric acid produces **nitrate** salts when NEUTRALISED, but...

Nitric acid reacts fine with alkalis, to produce nitrates, but it can play silly devils with metals and produce nitrogen oxides instead, so we'll ignore it here.

Revision of Acids and Metals — easy as squeaky pop...

Actually, this stuff isn't too bad I don't think. I mean it's fairly interesting. Not quite in the same league as base jumping, I grant you, but for Chemistry it's not bad at all. At least there are bubbles and flames and noise and that kind of thing. Anyway, <u>learn it, scribble it down, etc</u>...

Acids with Oxides and Hydroxides

Metal *Oxides* and Metal *Hydroxides* are *Alkalis*

1) Some metal oxides and metal hydroxides dissolve in water to produce alkaline solutions.

2) In other words, metal oxides and metal hydroxides are generally alkalis.

3) This means they'll react with acids to form a salt and water.

4) Even those that won't dissolve in water will still react with acid.

$$\textbf{Acid} \; + \; \textbf{Metal Oxide} \; \rightarrow \; \textbf{Salt} \; + \; \textbf{Water}$$

$$\textbf{Acid} \; + \; \textbf{Metal Hydroxide} \; \rightarrow \; \textbf{Salt} \; + \; \textbf{Water}$$

(These are neutralisation reactions of course.
You can use an indicator to tell when they've reacted completely.)

The *Combination* of Metal and Acid decides the *Salt*

This isn't exactly exciting but it's pretty easy, so try and get the hang of it:

Hydrochloric acid +	Copper oxide	→	Copper chloride	+	water
Hydrochloric acid +	Sodium hydroxide	→	Sodium chloride	+	water
Sulphuric acid +	Zinc oxide	→	Zinc sulphate	+	water
Sulphuric acid +	Calcium hydroxide	→	Calcium sulphate	+	water
Nitric acid +	Magnesium oxide	→	Magnesium nitrate	+	water
Nitric acid +	Potassium hydroxide	→	Potassium nitrate	+	water

The symbol equations are all pretty much the same.
Here's two of them:

$$H_2SO_4 \; + \; ZnO \; \rightarrow \; ZnSO_4 \; + \; H_2O$$
$$HNO_3 \; + \; KOH \; \rightarrow \; KNO_3 \; + \; H_2O$$

More big, important reactions to learn

These really are the meat and drink of GCSE chemistry. It's not that hard, it's just a matter of sitting down and learning the page. Once you've done that, cover the right side of the table above, and try to fill in the products of the reactions. Then do it again, covering the left side and filling in the reactants.

Non-metal Oxides

The **Oxides** of **non-metals** are usually **Acidic**, not alkaline

1) The best examples are the <u>oxides</u> of these three non-metals: <u>carbon</u>, <u>sulphur</u> and <u>nitrogen</u>.

2) <u>Carbon dioxide</u> dissolves in water to form <u>carbonic acid</u> which is a <u>weak</u> acid.

3) <u>Sulphur dioxide</u> combines with water and O_2 to form <u>sulphuric acid</u> which is a <u>strong</u> acid.

4) <u>Nitrogen dioxide</u> dissolves in water to form <u>nitric acid</u> which is a <u>strong</u> acid.

5) These three are all present in <u>acid rain</u> of course.

6) The <u>carbonic acid</u> is present in rain <u>anyway</u>, so even <u>ordinary</u> rain is slightly acidic.

CARBONIC ACID

Remember the three examples:

Non-metal oxides are acidic:

Carbon dioxide

Sulphur dioxide

Nitrogen dioxide

Acids are really dull — learn about them all the same

You've got to be a pretty serious career chemist to find this stuff interesting. Normal people (like you and me) just have to grin and bear it. Oh, and <u>learn it</u> as well, of course — don't forget the small matter of those little Exams you've got coming up...

Acids with Carbonates

More gripping reactions involving acids. At least there are some bubbles involved here.

Acid + Carbonate → Salt + Water + Carbon dioxide

Acid + Hydrogencarbonate → Salt + Water + Carbon dioxide

1) <u>Definitely</u> learn the fact that <u>carbonates</u> and <u>hydrogencarbonates</u> give off <u>carbon dioxide</u>.

2) If you also <u>practise</u> writing the following equations out <u>from memory</u>, it'll do you no harm at all.

hydrochloric acid + sodium carbonate → sodium chloride + water + carbon dioxide

$$2HCl \quad + \quad Na_2CO_3 \quad → \quad 2NaCl \quad + \quad H_2O \quad + \quad CO_2$$

hydrochloric acid + sodium hydrogencarbonate → sodium chloride + water + carbon dioxide

$$HCl \quad + \quad NaHCO_3 \quad → \quad NaCl \quad + \quad H_2O \quad + \quad CO_2$$

The Test For Carbon Dioxide: It turns limewater milky

Bubble the gas through <u>limewater</u>. If it's <u>carbon dioxide</u> the <u>limewater turns milky</u>.

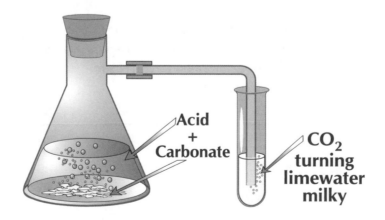

Acid
+
Carbonate

CO_2 turning limewater milky

Acids with Carbonates produce Salt, Water and CO_2

Another page of really, really useful stuff to learn. Make sure you memorise the equations really well, and be aware of the overlap between this hard-core chemistry, and Section Four on the environment and acid rain. Oh, and you'll need the test for CO_2 as well. Just learn it all.

Acids with Ammonia

Acids also react with ammonia...

Acids with Ammonia

| Dilute Acid + Ammonia → Ammonium salt |

Here are three of the most important acids reacting with ammonia:

Hydrochloric acid + Ammonia → Ammonium chloride

$$HCl_{(aq)} + NH_{3(aq)} \rightarrow NH_4Cl_{(aq)}$$

Sulphuric acid + Ammonia → Ammonium sulphate

$$H_2SO_{4(aq)} + 2NH_{3(aq)} \rightarrow (NH_4)_2SO_{4(aq)}$$

Nitric acid + Ammonia → Ammonium nitrate

$$HNO_{3(aq)} + NH_{3(aq)} \rightarrow NH_4NO_{3(aq)}$$

This last reaction, with nitric acid, produces the famous __ammonium nitrate__ fertiliser, much appreciated for its __double dose__ of essential nitrogen. (See page 54.)

Know your acids, or you'll get burned in the Exam

That's the last page on acids, thank goodness. <u>Learn</u> the last of these dreary facts and try to <u>scribble them down</u>. Pay special attention to those equations of course. Acids may be dull but they're still vitally important to your exam chances. Learn them well.

Warm-Up and Worked Exam Questions

To get the marks for questions on acids and alkalis, you need to have learnt the common reactions.
Take care when writing out equations to make sure that they balance, or you'll lose marks.

Warm-up Questions

1) What colour is universal indicator in strong acid?
2) Washing up liquid has a pH of 8. Is this an acid or alkali?
3) Indigestion tablets contain magnesium oxide. Is this an acid or alkali?
4) What ions are always formed when an acid is added to water?
5) Acid + Hydroxide → Salt + What?
6) Acid + Carbonate → Salt + What?

Worked Exam Question

There'll almost always be a question about acid reactions in the Exam.
Often it'll be as part of a rate of reaction question, but you still have to know the reaction.

1   This question is about the reaction of metals with acids.

a) What acid reacts with metals to produce sulphates?

   Sulphuric acid.

   (1 mark)

b) 0.5g of magnesium is added to a test tube of dilute hydrochloric acid.
   Bubbles of a gas are produced.

   i) State what gas is produced, and describe a test for this gas. *The classic squeaky pop test.*

      Hydrogen. A burning splint makes hydrogen ignite

      with a squeaky pop

      (2 marks)

   ii) Write a balanced equation for the reaction between hydrochloric acid and
       magnesium.

      2HCl + Mg → MgCl$_2$ + H$_2$

      Make sure it's balanced!

      (2 marks)

   iii) 0.5g of zinc is added to a test tube of dilute hydrochloric acid. Bubbles of gas are
        produced more slowly than when magnesium is added. Explain why this is.

      Zinc is less reactive than magnesium, so it reacts more

      slowly.

      (2 marks)

   iv) Would you expect any bubbles of gas to be produced if 0.5g of copper was added
       to a test tube of dilute hydrochloric acid? Explain your answer.

      No. Copper is less reactive than hydrogen.

      (2 marks)

Worked Exam Question

2   This question is about acids, alkalis and pH.

a)   Acids and alkalis ionise in aqueous solution.

i)   What ion is always present in an aqueous solution of acid?

H+

(1 mark)

ii)   What ion is always present in an aqueous solution of alkali?

OH⁻

(1 mark)

iii)   A general neutralisation reaction can be represented by the following word equation:

acid + alkali → salt + water.

Complete the following general ionic equation for a neutralisation reaction.

H+ + *OH⁻* → H_2O *H+ from acid and OH⁻ from alkali*

(2 marks)

b)   Universal indicator was used to measure the pH of several different substances:

Substance **A**	Indicator went blue-green
Substance **B**	Indicator went orange
Substance **C**	Indicator went red
Substance **D**	Indicator went indigo

i)   Which substance is a weak acid?

Substance B   *Orange means a weaker acid than red.*

(1 mark)

ii)   Which substance is ammonia?

Substance D   *Indigo means high pH, which means strong alkali.*

(1 mark)

iii)   Substance C is naturally present in the body. Suggest what substance C might be?

Stomach acid/hydrochloric acid   *Stomach acid has a very low pH*

(1 mark)

c)   The soil in Fred's garden has a pH of 5.4. What ionic compound could he add to the soil to make it neutral?

Slaked lime/calcium hydroxide $Ca(OH)_2$ / limestone $CaCO_3$

pH 5.4 is acidic. He needs lime (alkali) to make it neutral.

(1 mark)

Exam Questions

1   This question is about oxides of metals and non-metals.

   a)   Some metal oxides dissolve in water. Is the solution formed acidic or alkaline?

....

(1 mark)

   b)   For each of the following pairs of chemicals, write down the balanced equation if they react. If they do not react, explain why not.

      i)   Nitric acid and magnesium oxide.

....

(2 marks)

      ii)   Hydrochloric acid and calcium oxide.

....

(2 marks)

      iii)   Ammonia and sodium oxide.

....

(2 marks)

      iv)   Sulphuric acid and copper(II) oxide.

....

(2 marks)

   c)   Some non-metal oxides dissolve in water.

      i)   Carbon dioxide dissolves in water. Roughly what colour will this solution turn universal indicator paper?

....

(1 mark)

      ii)   Sulphur dioxide dissolves in water. Roughly what colour will this solution turn universal indicator paper?

....

(1 mark)

      iii)   Sulphur and nitrogen oxides are released into the air by the burning of fossil fuels. Describe briefly the environmental problems caused by these oxides when they dissolve in water vapour in the air.

....

....

....

(3 marks)

Exam Questions

2   Marble and limestone react with acid.

   a)   Marble chips are added to a flask containing hydrochloric acid. The acid fizzes, and the
        gas which is given off is passed through a test tube.

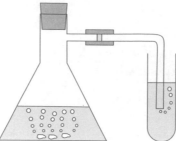

      i)   What is the main chemical in marble?

           ....
           (1 mark)

      ii) Complete and balance this equation for the reaction between marble and hydrochloric
           acid.

           .... + ...HCl → .... + H_2O + ...
           (3 marks)

      iii) The test tube in the diagram contains limewater. What happens to the limewater as
           the reaction progresses?

           ....

           ....
           (1 mark)

   b)   Write down a balanced equation for the reaction between sodium hydrogencarbonate and
        hydrochloric acid.

        ....
        (2 marks)

3   Ammonia is strongly alkaline, and reacts with acids.

   a)   What is the product of the reaction between ammonia and dilute hydrochloric acid?

        ....
        (1 mark)

   b)   i)   What is the most common use of ammonium nitrate?

           ....
           (1 mark)

      ii) Explain why ammonium nitrate is more suitable for this purpose than ammonium
           chloride.

           ....
           (1 mark)

Metals

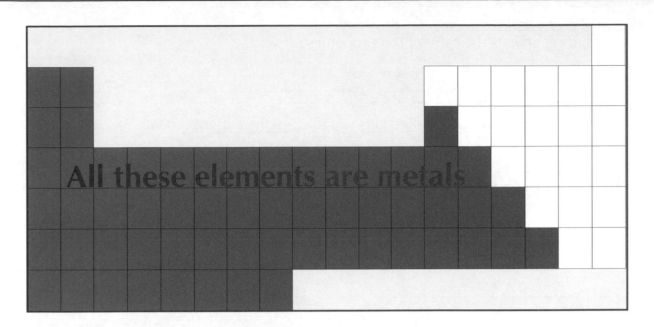

All these elements are metals

The Metallic Crystal Structure

1) <u>All metals</u> have the <u>same basic properties</u>.

2) These are due to the <u>special type of bonding</u> that exists in metals.

3) Metals consist of a <u>giant structure</u> of atoms held together with <u>metallic bonds</u>.

4) These special bonds allow the <u>outer electron(s)</u> of each atom to <u>move freely</u>.

5) This creates a "<u>sea</u>" of <u>free electrons</u> throughout the metal, which is what gives rise to many of the properties of metals.

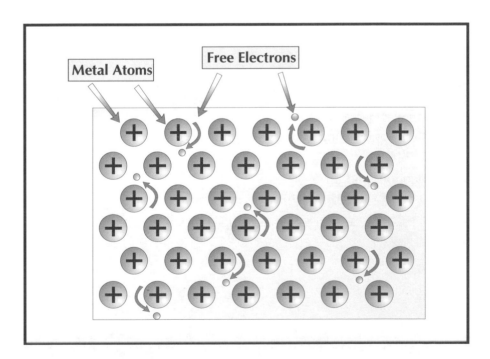

Metal Atoms   Free Electrons

Don't be left adrift on the electron 'sea'

This is just the same stuff which we covered in Section One, only now we're going to tell you about all the exciting implications of metallic bonding. I see you quiver with antici...   ...pation.

Metals

The 'sea' of free electrons in metallically bonded substances, gives metals their distinctive properties.

1) They all **conduct electricity**

This is entirely due to the free electrons which carry the current.

2) They're all **good conductors of heat**

Again this is entirely due to the free electrons which carry the heat energy through the metal.

3) Metals are **strong**, but also bendy and **malleable**

They are strong (hard to break), but they can be bent or hammered into different shapes.

4) They're all **shiny** (when freshly cut or polished)

5) They have **high melting and boiling points**

Which means you have to get them pretty hot to melt them (except mercury). E.g. copper 1100°C, tungsten 3377°C.

6) They can be mixed together to form **many useful alloys**:

1) Steel is an alloy (mixture) of iron and about 1% carbon. Steel is much less brittle than iron.

2) Bronze is an alloy of copper and tin. It's harder than copper but still easily shaped.

3) Copper and nickel (75%:25%) are used to make cupro-nickel which is hard enough for coins.

No wonder metals are so useful

Metals are really interesting. OK, well they're more interesting than saw dust. Anyway, make sure you know these SIX properties of metals. You should know the best way to do that by now. Cover up the page and write a mini-essay on why metals are so useful.

Non-Metals

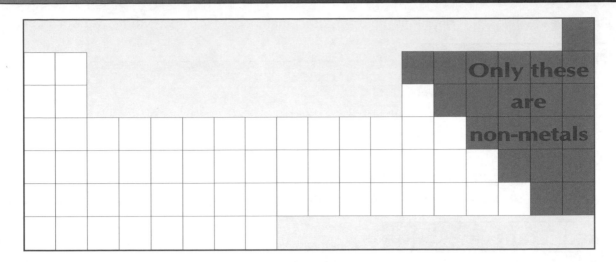

Non-Metal elements are either *dull, brittle solids* or they're *gases*

1) Only about <u>a quarter</u> of the elements are <u>non-metals</u>.

2) <u>Half</u> the non-metals are <u>gases</u> and half are <u>solids</u>.

3) <u>Bromine</u> is the only <u>liquid non-metal element</u>.
(Mercury is the only other element which is liquid
— at room temperature, that is.)

1) Non-metals are *poor conductors of heat*

2) Non-metals *don't conduct electricity* at all

— except for <u>graphite</u> which conducts because it has some
<u>free electrons</u> between the <u>layers</u> of its crystal structure.

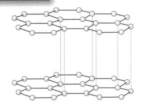

3) Non-metals usually bond in *small molecules*, e.g. O_2, N_2 etc.

4) But carbon forms *giant structures*:

Graphite
(pure carbon)

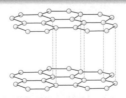

Diamond
(pure carbon)

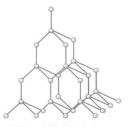

Non-metals have different properties

Metals and non-metals are really the only things that make Chemistry difficult. If it wasn't for them, the whole subject would be much more straightforward. Make sure you know what the differences are, and how the different properties relate to the different structures of metals and non-metals.

Warm-Up and Worked Exam Questions

You'd be wise to give these questions a proper go. Remember to look up anything you don't know.

Warm-up Questions

1) Is silicon a metal or a non-metal?
2) Is hydrogen a metal or a non-metal?
3) Bronze is an alloy of which two metals?
4) Which are better conductors of heat, metals or non-metals?
5) Are pure metals found in small molecules?

Worked Exam Question

This worked question should give you a good idea of what to write about metallic bonding.

1   This question is about the structure and properties of metals.

   a)   How many electrons are there in the outer shell of aluminium?

   3

   (1 mark)

   b)   Describe the crystal structure of metals and metallic bonding.

   Metals form giant crystal structures. The electrons in the outer shell of each atom are free to move around, forming a 'sea' of free electrons. The attraction between free electrons and metal ions bonds them together.

   Free-moving electrons are the key to metallic bonding.

   (3 marks)

   c)   Metals are good conductors of electricity. Explain why, in terms of their structure.

   The free electrons can move throughout the piece of metal. These moving electrons mean that an electric current can be passed through it.

   (2 marks)

   d)   What physical properties of copper make it suitable for making water pipes?

   Copper is strong, ductile and malleable. It doesn't corrode

   (2 marks)

   e)   What is an alloy?

   An alloy is a mixture of two or more different metals.

   (1 mark)

Exam Questions

1 a) Metals have a giant crystal structure with freely moving outer electrons.

   i) Explain how the structure of metals allows them to conduct heat.

   ....

   ....
   (2 marks)

   ii) The attraction between free electrons and metal ions in the metallic crystal
        structure is strong. How does this affect the melting points of metals?

   ....

   ....

   ....
   (3 marks)

   b) i) Describe the bonding in a molecule of oxygen.

   ....

   ....
   (2 marks)

   ii) Describe the bonding in diamond.

   ....

   ....
   (2 marks)

   c) This diagram shows one of the crystal structures of carbon.

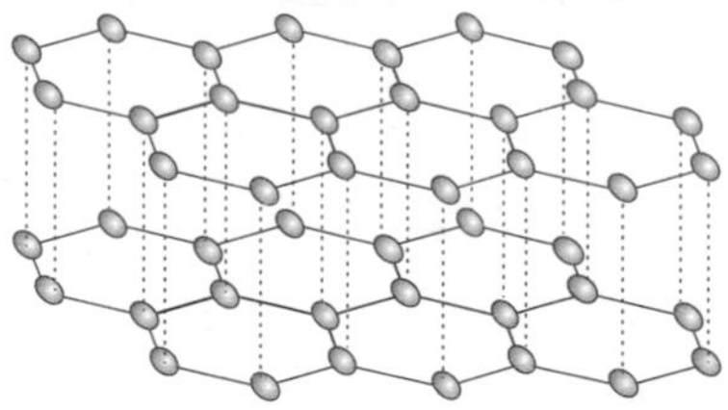

   i) What kind of pure carbon is shown in the diagram?

   ....
   (1 mark)

   ii) The type of carbon shown in the diagram can conduct electricity.
        Explain why, in terms of its crystal structure.

   ....

   ....

   ....
   (3 marks)

The Reactivity Series of Metals

You must learn this Reactivity Series

You really should know which are the more reactive metals and which are the less reactive ones.

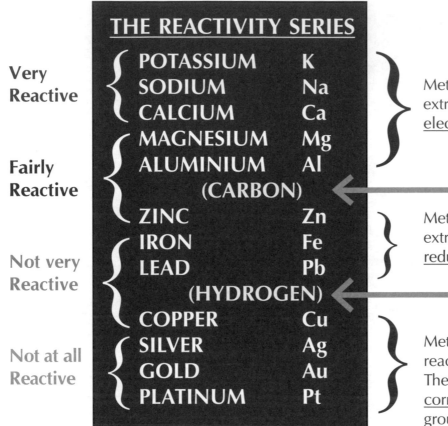

THE REACTIVITY SERIES

Very Reactive	POTASSIUM	K
	SODIUM	Na
	CALCIUM	Ca
Fairly Reactive	MAGNESIUM	Mg
	ALUMINIUM	Al
	(CARBON)	
Not very Reactive	ZINC	Zn
	IRON	Fe
	LEAD	Pb
	(HYDROGEN)	
	COPPER	Cu
Not at all Reactive	SILVER	Ag
	GOLD	Au
	PLATINUM	Pt

Metals <u>above carbon</u> must be extracted from their ores by <u>electrolysis</u>.

Metals <u>below carbon</u> can be extracted from their ores by <u>reduction</u> with <u>coke or charcoal</u>.

Metals <u>below hydrogen</u> don't react with <u>water</u> or <u>acid</u>. They don't easily <u>tarnish</u> or <u>corrode</u>, and are found in the ground in their pure state.

This <u>reactivity series</u> was determined by doing <u>experiments</u> to see <u>how strongly</u> metals <u>react</u>.

The <u>standard reaction</u> to determine reactivity is with <u>water</u>.

Learn all the details of this reaction on the next page.

The Reactivity Series is really important

The Examiners may ask you to predict what will happen when a certain metal is mixed with water or acid. You'll need to know the reactivity series and all the points about it if you want to answer these.

The Reactivity Series of Metals

Make sure you know about this experiment in reasonable detail:

Reacting Metals With **Water**

1) If a metal reacts with water it will always release hydrogen.

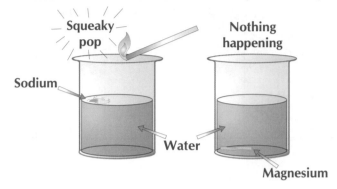

2) The more reactive metals react with cold water to form hydroxides:

SODIUM + WATER → SODIUM + HYDROGEN
HYDROXIDE

$2Na + 2H_2O \rightarrow 2NaOH + H_2$

3) The less reactive metals don't react quickly with water but **will** react with steam to form oxides:

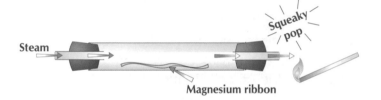

MAGNESIUM + WATER → MAGNESIUM OXIDE + HYDROGEN

$Mg + H_2O \rightarrow MgO + H_2$

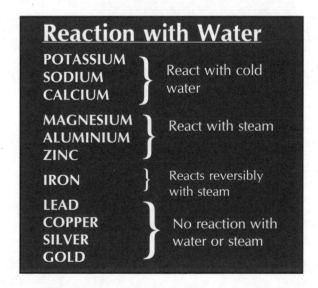

Reaction with Water

POTASSIUM	React with cold
SODIUM	water
CALCIUM	

MAGNESIUM
ALUMINIUM — React with steam
ZINC

IRON — Reacts reversibly with steam

LEAD
COPPER — No reaction with
SILVER — water or steam
GOLD

Reactivity can be determined by the reaction with water

Believe it or not they could easily give you a question asking if copper's more reactive than lead, or what happens when calcium is heated in water. That means all these details need learning.

Transition Metals

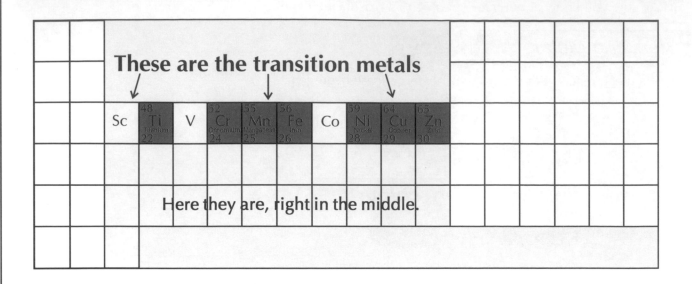

These are the transition metals

Here they are, right in the middle.

Titanium, Chromium, Manganese, Iron, Nickel, Copper, Zinc

You need to know the ones shown in red fairly well. But if they wanted to be mean in the Exam they could cheerfully mention one of the others — like scandium or cobalt or vanadium.

Don't let it get to you. They'll just be testing how well you can *"apply scientific knowledge to new information"*. In other words, just assume these "new" transition metals have all the properties you've already learnt for the others. That's all it means, but it can really worry some folk.

Transition Metals all have high melting points and high density

They're <u>typical metals</u>. They have the properties you would expect of a proper metal:

1) <u>Good conductors</u> of heat and electricity.
   They're very <u>dense</u>, <u>strong</u> and <u>shiny</u>.

2) Iron melts at 1500°C, copper melts at 1100°C and zinc melts at 400°C.

Learn the properties of transition metals

Now these are proper metals — hard, strong and shiny. The kind of stuff you can make cars and bridges and trains out of. Learn what they can do on these two pages.

Transition Metals

Transition Metals and their compounds all make good catalysts

1) Iron is the catalyst used in the Haber process for making ammonia.
2) Manganese(IV) oxide is a good catalyst for the decomposition of hydrogen peroxide.
3) Nickel is useful for turning oils into fats for making margarine.

The compounds are very colourful

1) The compounds are colourful due to the transition metal ion they contain.
   e.g. Potassium chromate(VI) is yellow. Potassium manganate(VII) is purple.
   Copper(II) sulphate is blue.

2) The colour of people's hair and also the colours in gemstones, like blue
   sapphires and green emeralds, and the colours in pottery glazes are all due to
   transition metals.
   ...And weathered (oxidised) copper is a lovely colourful green.

Transition Metals often have more than one ion, e.g. Fe^{2+}, Fe^{3+}

Two other examples are copper: Cu^+ and Cu^{2+} and chromium: Cr^{2+} and Cr^{3+}
The different ions usually have different colours too:

Fe^{2+} ions usually give green compounds.

Fe^{3+} ions are usually red/brown (e.g. rust).

Transition metals have many uses

1) Iron is used for manhole covers. Pure iron is very brittle, unlike steel
   which is more useful.
2) Copper is used for electric wiring and household water pipes. Copper and
   nickel combine to make cupronickel, which is used to make coins.
3) Zinc is used for galvanising iron. Zinc and copper make the alloy brass for
   trumpets and tubas.
4) Titanium is used to make strong, light alloys for aircraft and missiles.

Lots of pretty colours — that's what we like to see

There's quite a few things to learn about transition metals. First try to remember the headings on both
pages. Then learn the details that go under each one. Keep trying to scribble it all down.

Warm-Up and Worked Exam Questions

It really pays to learn the reactivity series. Knowing which metals are more reactive and which are less reactive makes these questions a lot easier.

Warm-up Questions

1) Does calcium react with cold water?
2) Does aluminium react with cold water? Does it react with water at all?
3) Which is more reactive, aluminium or lead?
4) Is magnesium more reactive than carbon?
5) Which two transition metals are alloyed together to make brass?
6) Is it possible for a transition metal to form different ions, say a 2+ ion and a 3+ ion?

Worked Exam Question

Exam questions on reactivity of metals tend to have some fairly easy parts in them. If you've learnt which are the reactive metals and which are the unreactive metals it won't be too hard.

1   Some metals are more reactive than others.

a)   Place these metals in order of reactivity (start with the most reactive):

Silver, Zinc, Potassium

Potassium, Zinc, Silver   *Potassium = very reactive, silver = unreactive*

(1 mark)

b)   Some metals react with water.

i)   What are the products of the reaction between sodium and water?

Sodium hydroxide and   *You should know this from work on Group I metals.*

hydrogen

(2 marks)

ii)   What are the products of the reaction between aluminium and steam?

Aluminium oxide   *Less reactive metals make an OXIDE, not a hydroxide.*

and hydrogen

(2 marks)

iii)   Describe what happens when aluminium is dropped into cold water.

Nothing   *Aluminium won't react with cold water.*

(1 mark)

iv)   Describe what you would see when hot steam is passed over aluminium in a tube. How would you identify the gas given off?

White aluminium oxide is formed. The gas produced will

ignite with a squeaky pop.   *The squeaky pop is the standard test for H_2.*

(2 marks)

Exam Questions

1   This table shows the reactivity series.

Potassium Sodium Calcium Magnesium Aluminium Carbon Zinc Iron Lead Hydrogen Copper Silver Gold

more reactive ➝ **less reactive**

a)   Which metals in the table are found in the ground in their pure state?

....
(1 mark)

b)   Which metals can't be reduced by carbon, and must be extracted by electrolysis?

....
(1 mark)

c)   Use the reactivity series to explain why silver won't react with water.

....

....
(2 marks)

2   This question is about the transition metals.

a)   Would you expect the transition metals to be good conductors of electricity? Explain why.

....

....
(2 marks)

b)   A catalyst is a substance that speeds up a chemical reaction without being used up in the reaction. Transition metals are used as catalysts in some industrial reactions. Which transition metal is used as a catalyst when hydrogenating vegetable oils to make margarine?

....
(1 mark)

c)   Transition metals lose varying numbers of outer shell electrons to form stable ions.

   i)   Copper forms a copper(II) ion, Cu^{2+}. What other ion can it form?

....
(1 mark)

   ii)   Manganese can form a manganate(VII) ion, MnO_4^-. What colour is potassium manganate(VII), $KMnO_4$?

....
(1 mark)

Revision Summary

There's some serious Chemistry in Section Five.
I suppose it makes up for Section Four being so easy. This is where all the really tough stuff is.
All I can say is, just keep trying to learn it. These questions will give you some idea of how well
you're doing. For any you can't do, you'll find the answers somewhere in Section Five.

1) What two properties did chemists base the early periodic table on?

2) Who had the best try at it and why was his table so clever?

3) What feature of atoms determines the order of the elements in the modern Periodic Table?

4) What are the Periods and Groups? Explain their significance in terms of electrons.

5) Draw diagrams to show the electron arrangements for the first twenty elements.

6) Explain the trend in reactivity of metals and non-metals as you go down a Group, using the notion of "shielding".

7) What are the electron arrangements of the noble gases? What are their properties?

8) Give two uses each for helium, neon and argon.

9) Which Group contains the alkali metals? What is their outer shell like?

10) List four physical properties, and two chemical properties of the alkali metals.

11) Give details of the reactions of the alkali metals with water.

12) What can you say about the pH of alkali metal oxides and hydroxides?

13) Describe the trends in appearance and reactivity of the halogens as you go down the Group.

14) List four properties common to all the halogens.

15) Give details, with equations, of the reaction of the halogens with metals.

16) Give details, with equations, of the displacement reactions of the halogens.

17) What is hydrogen chloride? Exactly how do you produce an acid solution from it?

18) What are the two sources of salt and what are the three main uses of it?

19) Draw a full diagram of the electrolysis of salt and list the three useful products it creates.

20) Give uses for the three products from the electrolysis of brine.

21) Give a use for each of the four halogens: fluorine, chlorine, bromine and iodine.

22) Describe fully the colour of universal indicator for every pH value from 1 to 14.

23) What type of ions are always present in a) acids and   b) alkalis? What is neutralisation?

24) What is the equation for reacting an acid with a metal? Which metal(s) don't react with acid?

25) What type of salts do hydrochloric acid and sulphuric acid produce?

26) What type of reaction is "acid + metal oxide" or "acid + metal hydroxide"?

27) What about the oxides of non-metals — are they acidic or alkaline?

28) What are the equations for reacting acids with carbonates and hydrogencarbonates?

29) What is the equation for reacting dilute acid with ammonia?

30) What proportion of the elements are metals? What do all metals contain?

31) List six properties of metals. List four properties of non-metals.

32) Write down twelve common metals in the order of the Reactivity Series.

33) Where do carbon and hydrogen fit in and what is the significance of their positions?

34) Describe the reaction of all twelve metals with water (or steam).

35) List four properties of transition metals, and two properties of their compounds.

36) Name six transition metals, and give uses for three of them.

Rates of Reaction

Reactions can go at all sorts of different *rates*

1) One of the <u>slowest</u> is the <u>rusting</u> of iron (it's not slow enough though —
   what about my little MG).
2) Other slow reactions include <u>chemical weathering</u>, like acid rain damage
   to limestone buildings.
3) A <u>moderate speed</u> reaction is a <u>metal</u> (like magnesium) reacting with <u>acid</u>
   to produce a <u>gentle stream of bubbles</u>.
4) A <u>really fast</u> reaction is an <u>explosion</u>, where it's all over in a <u>fraction of a second</u>.

Three ways to Measure the *Speed* of a Reaction

The <u>speed of reaction</u> can be observed <u>either</u> by how quickly the <u>reactants are used up</u> or
how quickly the <u>products are forming</u>. It's usually a lot easier to measure <u>products forming</u>.

There are <u>three different ways</u> that the speed of a reaction can be <u>measured</u>:

1) *Precipitation*

This is when the <u>product</u> of the reaction is a
<u>precipitate</u> which <u>clouds the solution</u>.
Observe a <u>marker</u> through the solution and measure
<u>how long it takes</u> for it to <u>disappear</u>.

2) *Change in mass* (usually gas given off)

Any reaction that <u>produces a gas</u> can be carried out on a <u>mass balance</u>
and as the gas is released the mass <u>disappearing</u> is easily measured.

3) The *volume* of gas given off

This involves the use of a <u>gas syringe</u> to measure the
volume of gas given off. That's about all there is to it.

You might need this for calculations

In the Exam, you might get a question which describes an experiment and asks you to do calculations
about the reaction rate from the data collected by the above methods.

Rates of Reaction

The Rate of a *Reaction Depends* on *Four Things*:

1) TEMPERATURE

2) CONCENTRATION — (or PRESSURE for gases)

3) CATALYST

4) SIZE OF PARTICLES — (or SURFACE AREA)

Learn them

Typical Graphs for Rate of Reaction

The plot below shows how the speed of a particular reaction varies under *different conditions*. The quickest reaction is shown by the line that becomes *flat* in the *least* time.

1) **Graph 1** represents the original <u>fairly slow</u> reaction.

2) **Graphs 2 and 3** represent the reaction taking place <u>quicker</u> but with the <u>same initial amounts</u> of reactants.

3) The <u>increased rate</u> could be due to <u>any</u> of these:

   a) increase in <u>temperature</u>

   b) increase in <u>concentration</u> (or pressure)

   c) solid reactant crushed up into <u>smaller bits</u>

   d) <u>catalyst</u> added.

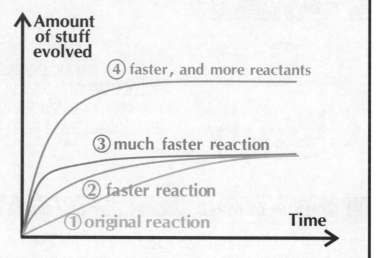

4) **Graph 4** produces <u>more product</u> as well as going <u>quicker</u>. This can <u>only</u> happen if <u>more reactant(s)</u> are added at the start.

Heat it up for a fast, furious reaction

There's lots of information on this page. To learn it all, you've just got to keep reading it over. The graph is especially important, but if you don't learn the notes that go with it, it won't do you much good at all. Practise by <u>covering the page</u>, sketching the graph and scribbling the notes.

Collision Theory

Reaction rates are explained perfectly by Collision Theory. It's really simple.

> It just says that the rate of a reaction simply depends on how often and how hard the reacting particles collide with each other.
>
> The basic idea is that particles have to collide in order to react, and they have to collide hard enough as well.

More Collisions increases the Rate of Reaction

All the methods of increasing the rate of reactions can be explained in terms of increasing the number of collisions between the reacting particles.

1) TEMPERATURE increases the number of collisions

When the temperature is increased the particles all move quicker. If they're moving quicker, they're going to have more collisions.

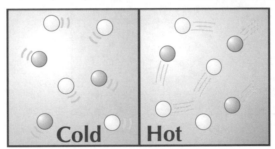

2) CONCENTRATION (or PRESSURE) increases the number of collisions

If the solution is made more concentrated it means there are more particles of reactant knocking about between the water molecules which makes collisions between the important particles more likely. In a gas, increasing the pressure means the molecules are more squashed up together so there are going to be more collisions.

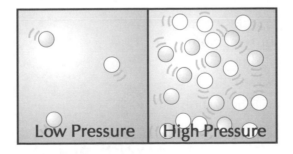

It's all about the frequency of collisions

Once you understand collision theory, the next two points are pretty obvious really. If you heat particles so they move faster, they're going to bump into each other more often. If you increase the number of particles whizzing round in a space, you'll get more collisions. Simple.

Collision Theory

3) SIZE OF SOLID PARTICLES (or SURFACE AREA) increases collisions

If one of the reactants is a <u>solid</u> then <u>breaking it up</u> into <u>smaller</u> pieces will <u>increase</u> <u>its surface area</u>. This means the particles around it in the solution will have <u>more</u> <u>area to work on</u> so there'll be <u>more useful collisions</u>.

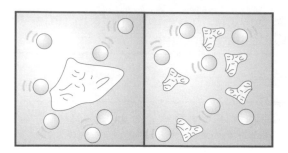

4) CATALYST increases the number of successful collisions

A <u>catalyst</u> works by giving the <u>reacting particles</u> a <u>surface</u> to <u>stick to</u> where they can <u>bump</u> into each other. This increases the <u>number of successful collisions</u>.

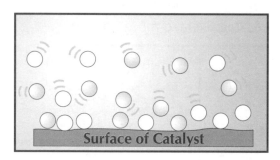

Surface of Catalyst

Faster Collisions increase the Rate of Reaction

<u>Higher temperature</u> also increases the <u>energy</u> of the collisions, because it makes all the particles <u>move faster</u>.

Faster collisions are ONLY caused by increasing the temperature

Reactions <u>only happen</u> if the particles collide with <u>enough energy</u>. At a <u>higher temperature</u> there will be <u>more particles</u> colliding with <u>enough energy</u> to make the reaction happen.

This <u>initial energy</u> is known as the <u>activation energy</u>, and it's needed to <u>break the initial bonds</u> (see page 179).

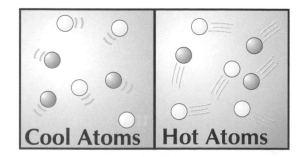

Cool Atoms | Hot Atoms

Collision Theory — it's always the other driver

This is quite easy I think. The more often particles collide and the harder they hit, the greater the reaction rate. There's a few extra picky details of course, <u>but you've got to LEARN them</u>...

Four Experiments on Rate of Reaction

REMEMBER: Any reaction can be used to investigate any of the four factors that affect the rate. These pages illustrate four important reactions, but only one factor has been considered for each. But we could just as easily use, say, the marble chips/acid reaction to test the effect of temperature instead.

1) Reaction of Hydrochloric Acid and Marble Chips

This experiment is often used to demonstrate the effect of breaking the solid up into small bits.

1) Measure the volume of gas evolved with a gas syringe and take readings at regular intervals.
2) Make a table of readings and plot them as a graph.
3) Repeat the experiment with exactly the same volume of acid, and exactly the same mass of marble chips, but with the marble more crunched up.
4) Then repeat with the same mass of powdered chalk instead of marble chips.

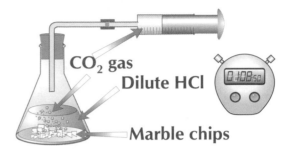

CO₂ gas
Dilute HCl
Marble chips

These graphs show the effect of using finer particles of solid

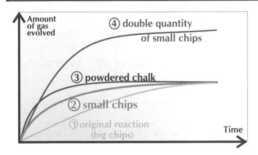

1) The increase in surface area causes more collisions so the rate of reaction is faster.
2) Graph 4 shows the reaction if a greater mass of small marble chips is added.
3) The extra surface area gives a quicker reaction and there is also more gas evolved overall.

2) Reaction of Magnesium Metal With Dilute HCl

1) This reaction is good for measuring the effects of increased concentration, (as is the marble/acid reaction).
2) This reaction gives off hydrogen gas, which we can measure with a mass balance, as shown. (The other method is to use a gas syringe, as above.)

These graphs show the effect of using more concentrated acid solutions

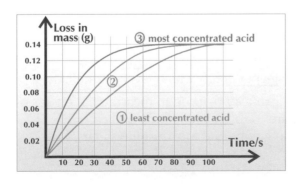

1) Take readings of the mass at regular time intervals.
2) Put the results in a table and work out the loss in mass for each reading. Plot a graph.
3) Repeat with more concentrated acid solutions but always with the same amount of magnesium.
4) The volume of acid must always be kept the same too — only the concentration is increased.
5) The three graphs show the same old pattern. Higher concentration gives a steeper graph with the reaction finishing much quicker.

Learn these examples

Of course, other reactions could be used to gather similar data on reaction rates, so don't bank on these exact experiments coming up. Make sure you know the theory so you can apply it to anything.

Four Experiments on Rate of Reaction

3) *Sodium Thiosulphate* and *HCl* produce a *Cloudy Precipitate*

1) These two chemicals are both <u>clear solutions</u>.
2) They react together to form a <u>yellow precipitate</u> of <u>sulphur</u>.
3) <u>The experiment</u> involves watching a black mark <u>disappear</u> through the <u>cloudy sulphur</u> and <u>timing</u> how long it takes to go.

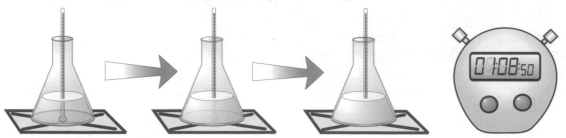

4) The reaction can be <u>repeated</u> for solutions at different <u>temperatures</u>.

5) The <u>depth</u> of liquid must be kept the <u>same</u> each time, of course.

6) The results will show that the <u>higher the temperature</u> the <u>quicker the reaction</u> and therefore the <u>less time</u> it takes for the mark to <u>disappear</u>.   These are typical results:

Temperature	20°C	25°C	30°C	35°C	40°C
Time taken for mark to disappear	193s	151s	112s	87s	52s

This reaction can <u>also</u> be used to test the effects of <u>concentration</u>.

One sad thing about this reaction is <u>it doesn't give a set of graphs</u>.
All you get is a set of <u>readings</u> of how long it took till the mark disappeared at each temperature.

4) *The Decomposition of Hydrogen Peroxide*

This is a <u>good</u> reaction for showing the effect of different <u>catalysts</u>.
The decomposition of hydrogen peroxide is:

$$2H_2O_2 \rightleftharpoons 2H_2O + O_2$$

1) This is normally <u>quite slow</u> but a sprinkle of <u>manganese(IV) oxide catalyst</u> speeds it up. Other catalysts which work are a) <u>potato peel</u> and b) <u>blood</u>.

2) <u>Oxygen gas</u> is given off which provides an <u>ideal way</u> to measure the rate of reaction using the good old <u>gas syringe</u> method.

O$_2$ gas
Hydrogen Peroxide
Catalyst

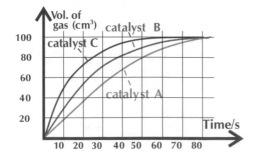

1) Same graphs of course.

2) <u>Better catalysts</u> give a <u>quicker reaction</u> which is shown by a <u>steeper graph</u> which levels off quickly.

3) This reaction can also be used to measure the effects of <u>temperature</u>, or of <u>concentration</u> of the H_2O_2 solution. The graphs will look <u>just the same</u>.

Four Top Rate Reactions to learn
There's always so much happening with reaction rates. Are we measuring gas, or mass, or cloudiness? Is it the effect of temp. or conc. or catalyst or surface area? You'll just have to work at it I'm afraid.

Warm-Up and Worked Exam Questions

The Examiners love to ask questions on rates of reaction. Better learn this section properly, hadn't you?
Try these easy starters before you go on to the main course.

Warm-up Questions

1) Write down four things that the rate of a reaction depends on.
2) Write down one way of making liquid particles move faster.
3) Particles can only react if they collide. Write down three ways of decreasing the number of collisions in one minute.
4) Name two things you can add to hydrogen peroxide to make it decompose faster.
5) A piece of zinc is dropped into a beaker of dilute acid. In the reaction, hydrogen gas is given off. Give two methods of measuring the reaction rate.

Worked Exam Questions

Remember to look through the worked examples really carefully before you go on to the exam questions.

1   Hannah investigated the rate of reaction of magnesium with hydrochloric acid.
     She used this apparatus:

a)   Complete this word equation for the reaction:

     magnesium + hydrochloric acid → <u>*magnesium chloride*</u> **+ hydrogen**

     (1 mark)

b)   Hannah did her experiment 3 times. Each time she used hydrochloric acid of a
     different concentration. She plotted this graph to show her results.

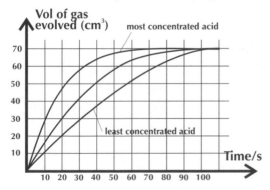

     i)   Describe what the graph shows about how the rate of reaction is affected by the
          concentration of the hydrochloric acid.

          Don't just write that the rate   <u>As higher concentrations of hydrochloric acid</u>
          increases – you must say what
          makes the rate increase!   <u>are used, the rate of the reaction increases.</u>

     (1 mark)

Worked Exam Questions

ii) Use collision theory to explain your answer to (i).

At higher acid concentrations there are more acid particles in a certain volume of solution. There will therefore be a greater number of effective collisions in a certain time.

(2 marks)

2   Marcus investigated the decomposition of hydrogen peroxide. He used this apparatus to find out how to make the rate of the reaction as fast as possible:

a)   Complete the symbol equation below, making sure that it is balanced.

$$2H_2O_2 \rightarrow \quad \underset{\text{...................................}}{2\ H_2O} \quad + \ O_2$$

(1 mark)

b)   Marcus tried out three catalysts. He plotted his results on a graph:

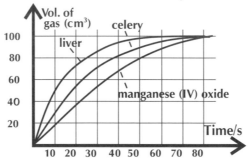

The quicker the volume of gas rises, the faster the reaction.

State which of the three catalysts is best and give a reason for your choice.

Liver is the best catalyst. I chose this because the graph for liver is steeper. The steeper the graph, the faster the reaction.

(3 marks)

c)   Write down two other changes Marcus could make to the reaction conditions so that the reaction is faster.

Marcus could do the reaction at a higher temperature or use a more concentrated hydrogen peroxide solution.

He could also have chopped the catalyst up more   *(2 marks)*

Exam Questions

1   Magnesium reacts with hydrochloric acid to make magnesium chloride solution
and hydrogen gas.

    a)   Draw a diagram of the apparatus you could use to measure the volume of hydrogen gas
produced every 30 seconds.

(2 marks)

    b)   Here are some results from this experiment:

Time in seconds	0	15	30	45	60	75
Volume of gas in cm^3	0	51	79	96	100	100

     i)   Plot the points on the graph.

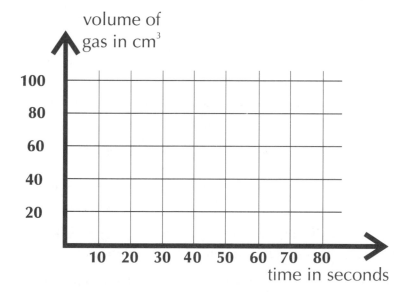

(2 marks)

     ii)   Draw a line of best fit.

(1 mark)

Exam Questions

2   Marble chips (calcium carbonate) react with hydrochloric acid like this:

calcium carbonate + hydrochloric acid → calcium chloride + water + carbon dioxide

Sarah investigated the rate of the reaction using this apparatus:

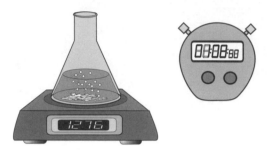

a)   What would you expect to happen to the mass reading during the experiment?

....
(1 mark)

b)   Sarah plotted a graph of her results:

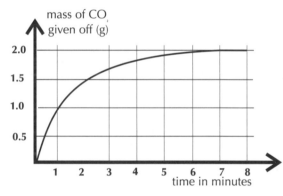

How many minutes did the reaction take to finish completely?

....
(1 mark)

c)   Sarah repeated the experiment with larger marble chips. She made sure she used the same mass of marble chips and the same amount of hydrochloric acid as in the first experiment.

     i)   Add a curve to the graph to show the progress of the reaction with the larger marble chips.

(2 marks)

     ii)   Explain in terms of particles why the rate of this reaction depends on the size of the marble chips.

....

....

....
(3 marks)

Catalysts

Many reactions can be <u>speeded up</u> by adding a <u>catalyst</u>.

> A <u>*CATALYST*</u> is a substance which <u>*INCREASES*</u>
> the speed of a reaction, without being
> <u>*CHANGED*</u> or <u>*USED UP*</u> in the reaction.

1) Catalysts *lower* the *Activation Energy*

1) Catalysts <u>lower</u> the <u>activation energy</u> (see page 179) of reactions, making it <u>easier</u> for them to happen.

2) This means a <u>lower temperature</u> can be used.

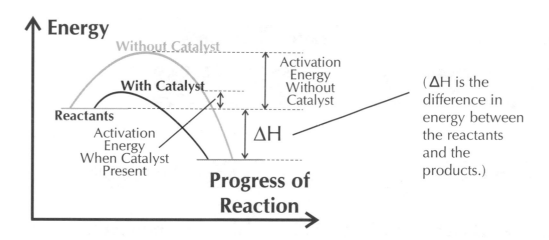

2) Catalysts *work best* when they have a *Big Surface Area*

1) Catalysts are usually used as a <u>powder</u> or <u>pellets</u> or a <u>fine gauze</u>.

2) This gives them <u>maximum surface area</u> to enable the reacting particles to <u>meet up</u> and do the business.

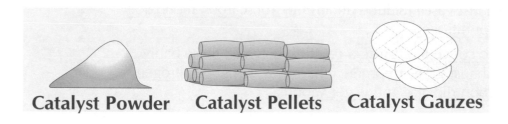

Catalyst Powder   **Catalyst Pellets**   **Catalyst Gauzes**

Some reactions need a catalyst to go at all

Remember, because catalysts lower the activation energy, they can make some reactions happen which would otherwise not occur, or need a lot of heating. They're used a lot in industry.

Catalysts

3) Catalysts Help **Reduce Costs** in Industrial Reactions

1) Catalysts increase the rate of many industrial reactions, which saves a lot of money simply because the plant doesn't need to operate for as long to produce the same amount of stuff.

2) Alternatively, a catalyst will allow the reaction to work at a much lower temperature and that can save a lot of money too. Catalysts are therefore very important for commercial reasons.

3) Catalysts are used over and over again. They may need cleaning but they don't get used up.

4) Different reactions use different catalysts.

5) Transition metals are common catalysts in many industrial reactions. Know these two:

a) An **Iron Catalyst** is used in the **Haber Process**

$$N_{2\,(g)} + 3H_{2\,(g)} \xrightarrow{\text{Iron Catalyst}} 2NH_{3\,(g)}$$

(See page 53 and page 172.)

b) A **Platinum Catalyst** is used in the production of **Nitric Acid**

Ammonia + Oxygen $\xrightarrow{\text{Platinum Catalyst}}$ Nitrogen monoxide + Water

The nitrogen monoxide is then reacted with oxygen and water to make nitric acid.

(See page 54.)

4) **Catalytic Converters** in Cars contain **Platinum**

1) These are fitted in the exhaust system of all new cars.

2) Normal exhaust gases include unburnt petrol, carbon monoxide and oxides of nitrogen.

3) The catalytic converters cause a reaction between these badly polluting exhaust gases to produce harmless gases: — nitrogen, oxygen, carbon dioxide and water vapour.

Catalysts are like great jokes — you can use them over and over...

Make sure you learn all the amazing things catalysts can do, and what people use them for. There are loads of examples of catalysts being used all through this book. Look up the page references above to check a couple of them out. I know there's a lot to learn here, but you need to know it all.

Biological Catalysts — Enzymes

Enzymes are Catalysts produced by Living Things

1) Living things have thousands of different chemical processes going on inside them.

2) The quicker these happen the better, and raising the temperature of the body is an important way to speed them up.

3) However, there's a limit to how far you can raise the temperature before cells start getting damaged, so living things also produce enzymes which act as catalysts to speed up all these chemical reactions without the need for high temperatures.

4) Enzymes themselves only perform well within a fairly narrow range of temperatures.

5) Every different biological process has its own enzyme designed especially for it.

6) For example the way an apple turns brown when cut is caused by a particular enzyme.

7) Chemists are now starting to use biological catalysts more and more for their own purposes.

8) Enzymes have many advantages over traditional non-organic catalysts:

   a) There's a huge variety of enzymes.

   b) They're not scarce like many metal catalysts e.g. platinum.

   c) They work best at low temperatures and pressures, which keeps costs down.

   d) They can be carefully selected to do a precise job.

EXAMPLES:
Enzymes are used in   "BIOLOGICAL" WASHING POWDERS

DISHWASHER POWDERS

TREATMENT OF LEATHER

Enzymes are really clever
Compared to enzymes, catalysts are stupid dull pieces of kit. Enzymes are so much better in many ways — they're specific to just one reaction, relatively inexpensive, and they don't need the kind of high temperatures and pressures that normal catalysts need. And our bodies make them. Clever us.

Biological Catalysts — Enzymes

Enzymes *Like it* Warm *but* Not Too Hot

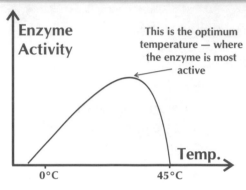

Enzyme Activity

This is the optimum temperature — where the enzyme is most active

Temp.

0°C   45°C

1) The chemical reactions in living cells are quite fast in conditions that are warm rather than hot.

2) This is because the cells use catalysts called enzymes, which are protein molecules.

3) Enzymes are usually damaged by temperatures above about 45°C, and as the graph shows, their activity drops off sharply when the temperature gets a little too high.

Enzymes *Like it the* Right pH *too*

The pH affects the activity of enzymes, in a similar way to temperature.

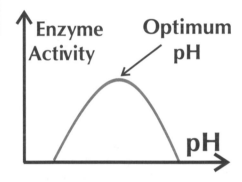

Enzyme Activity

Optimum pH

pH

Freezing *food* stops *the enzyme activity (and the bacteria)*

1) At lower temperatures, enzyme activity also drops quite quickly.

2) By 0°C there's virtually nothing happening.

3) This is the idea behind refrigeration, where foods are kept at about 4°C to keep enzyme and bacterial activity to a minimum so that food stays fresher for longer.

4) Freezers store food at about -20°C and at this temperature bacteria and enzymes don't function at all.

5) However, they're not destroyed by freezing and once the food thaws out they spring back into action. So frozen food should be thawed carefully and then cooked again before eating.

6) Cooking destroys all bacteria and enzymes, so properly cooked food is safe to eat.

7) However, even cooked foods will go off pretty rapidly if left in a warm place.

Enzymes are a bit delicate

This page is definitely a candidate for the mini-essay method. Write one about maximising enzyme performance. Scribble down the facts, then look back and see what you missed.

Uses of Enzymes

Living cells use chemical reactions to produce <u>new materials</u>. Many of these reactions provide products which are <u>useful</u> to us. Here are <u>three</u> important examples:

Yeast in Brewing of Beer and Wine: **Fermentation**

1) <u>Yeast cells</u> convert <u>sugar</u> into <u>carbon dioxide</u> and <u>alcohol</u>.

2) They do this using the <u>enzyme ZYMASE</u>.

3) The main thing is to <u>keep the temperature just right</u>.

4) If it's <u>too cold</u> the enzyme won't work very <u>quickly</u>.

5) If it's <u>too hot</u> it will <u>destroy</u> the enzyme.

6) This biological process is called <u>fermentation</u> and is used for making alcoholic drinks like <u>beer and wine</u>.

FERMENTATION is the process of *yeast* converting *sugar* into *carbon dioxide* and *alcohol*.

$$\text{Glucose} \xrightarrow{\text{Zymase}} \text{Carbon dioxide} + \text{Ethanol} \quad (+ \text{ Energy})$$

Yeast in Bread-making: **Fermentation again**

1) The reaction in <u>bread-making</u> is <u>exactly the same</u> as that in <u>brewing</u>.

2) Yeast cells use the enzyme <u>zymase</u> to break down sugar and this gives them <u>energy</u>.

3) It also releases carbon dioxide gas and alcohol as waste products.

4) The <u>carbon dioxide gas</u> is produced <u>throughout</u> the bread mixture and forms in <u>bubbles</u> everywhere.

5) This makes the bread <u>rise</u> and gives it its familiar texture. The small amount of alcohol also gives the bread some extra flavour, no doubt.

6) When the bread is put in the <u>oven</u> the yeast is <u>killed</u> and the <u>reaction stops</u>.

*Enzymes are used to do **things to foods***

1) The <u>proteins</u> in some <u>baby foods</u> are '<u>pre-digested</u>' using protein-digesting enzymes (proteases).

2) The <u>centres</u> of <u>chocolates</u> can be softened using enzymes.

3) Enzymes can turn <u>starch syrup</u> (yuk) into <u>sugar syrup</u> (yum).

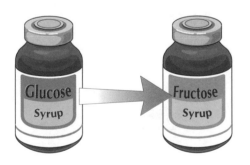

4) <u>Glucose syrup</u> can be turned into <u>fructose syrup</u> using enzymes. Fructose is <u>sweeter</u>, so you can use <u>less</u> of it — good for slimming foods and drinks.

Examples like these could turn an Examiner's head

Just look at these lovely examples to learn. They're interesting, involve food and have pretty pictures — they should be really easy to remember by anyone's standards. Learn and enjoy.

Uses of Enzymes

Yoghurt and *Cheese* making

1) Milk is mixed with <u>specially grown cultures</u> of bacteria.

2) This mixture is kept at the <u>ideal temperature</u> for the bacteria and their enzymes to work.

3) For <u>yoghurt</u> this is <u>pretty warm</u> at about <u>45°C</u>.

4) The <u>yoghurt-making bacteria</u> convert <u>lactose</u> (the natural sugar found in milk), into <u>lactic acid</u>. This gives yoghurts their slightly <u>sour</u> taste.

5) <u>Cheese</u> on the other hand matures better in <u>cooler conditions</u>.

6) <u>Various</u> bacterial enzymes can be used in <u>cheese making</u> to produce different <u>textures</u> and <u>tastes</u>.

Enzymes are used in **Biological Detergents**

1) <u>Enzymes</u> are the '<u>biological</u>' ingredients in biological detergents and washing powders.

2) They're mainly <u>protein-digesting</u> enzymes (proteases) and <u>fat-digesting</u> enzymes (lipases).

3) Because the enzymes attack <u>animal</u> and <u>plant</u> matter, they're ideal for removing <u>stains</u> like <u>food</u> or <u>blood</u>.

Using Enzymes in *Industry* takes a bit of fiddling

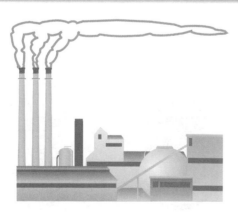

1) In an industrial process, the <u>temperature</u> and <u>pH</u> have to be right so the enzymes <u>aren't damaged</u>, and keep working for a long time.

2) The enzymes have to be kept from <u>washing away</u>. They can be mixed into <u>plastic beads</u>, or trapped in an <u>alginate bed</u> (seaweed mush).

3) Because enzymes work for a long time, you can <u>continually</u> pass chemicals over them to react, and tap off the product at the other end.

This page is easy, but that doesn't mean you can ignore it

I know this seems more like Food Technology than Chemistry, but you're expected to know all these details of making bread, wine, cheese, yoghurt, weird sugars and baby food, detergents, and the industrial bit. <u>Mini-essays again, I'd say.</u> <u>Enjoy.</u>

Warm-Up and Worked Exam Questions

Expect some kind of question about catalysts somewhere in the Exam.
You may have to give examples of catalysts in industrial chemistry or in food processing.

Warm-up Questions

1) What is an enzyme?
2) What is the maximum operating temperature of most enzymes?
3) What reaction is catalysed by the enzyme zymase?
4) Describe two methods used in industry to prevent enzymes washing away during processing.
5) Biological washing powders contain proteases and lipases. What kind of stains do these two enzymes digest?

Worked Exam Question

This worked example is just like the sort of question on catalysts that you could get in the Exam.

1   Look at these two diagrams which show the energy change during a reaction.

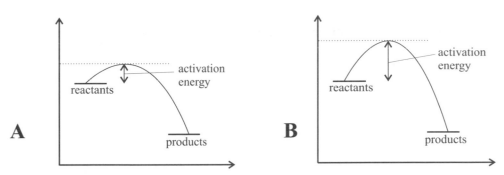

a) Which diagram shows the reaction taking place in the presence of a catalyst?
   Diagram A   The activation energy of a reaction is lowered by a catalyst.
                        Diagram A has a lower activation energy.
   (1 mark)

b) Catalysts are usually used in the form of a fine powder, or a fine mesh. Why is this?
   Think about collision theory...   *It gives them the maximum surface area.*
   Catalysts work best when they have a large surface area to
   "catch" reacting particles
   (2 marks)

c) Describe two ways in which catalysts reduce costs in industrial chemistry.
   Catalysts allow a reaction to take place at lower temperatures,
   which is cheaper as heating things costs money. Catalysts
   increase the rate of reaction, meaning that more product can get
   made in less time, which cuts costs.   You need to mention how each way cuts costs to get the marks.
   (4 marks)

Exam Questions

1 a) Beer is brewed from sugary barley malt. Alcohol is produced by fermentation.

      i) Write down a definition of fermentation.

        Fermentation is ....

        ....

(2 marks)

      ii) Write down a word equation for fermentation.

        ....

(2 marks)

      iii) Fermentation does not happen if the barley malt is too hot or too cold. Explain why.

        ....

        ....

(2 marks)

   b) Advertisements for biological detergents claim that they can digest stains in a 40°C wash.

      i) Would you expect a biological detergent to function as well at 60°C? Explain why.

        ....

        ....

(2 marks)

      ii) What kind of enzyme would be able to digest stains like egg and blood?

        ....

(1 mark)

2 This question is about enzymes in food preparation.

   a) Corn contains a lot of starch. Briefly describe how syrup containing a lot of glucose is made from corn.

      ....

      ....

(1 mark)

   b) Fructose is sweeter than sucrose. Explain why fructose is used in some "diet" or "healthy option" foods in place of sucrose.

      ....

(1 mark)

   c) What enzyme causes bread to rise?

      ....

(1 mark)

Exam Questions

3 a) i) What catalyst is used in the Haber Process?

....
(1 mark)

ii) What catalyst is used when making nitrogen monoxide (as part of nitric acid synthesis)?

....
(1 mark)

b) Describe the advantages of enzymes over traditional non-organic catalysts.

....

....

....

....
(4 marks)

4 This graph shows enzyme activity against temperature.

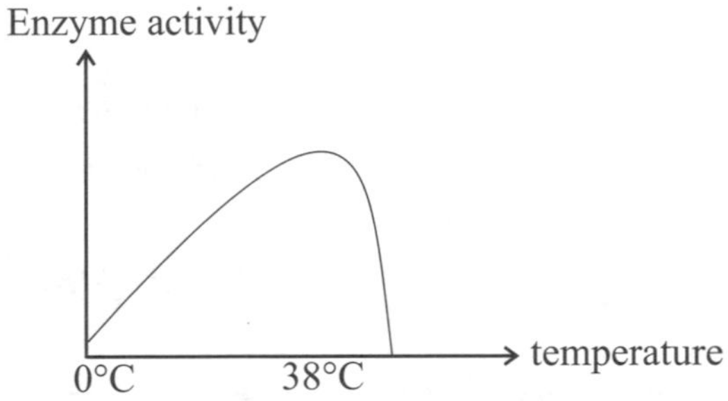

a) Explain why the enzyme activity drops off sharply over about 38°C.

....

....
(1 mark)

b) i) Explain in terms of bacterial enzyme activity why freezing food stops it going off.

....

....
(2 marks)

ii) Will frozen food go off once it is thawed? Explain your answer.

....

....

....
(2 marks)

Simple Reversible Reactions

A <u>reversible reaction</u> is one which can go <u>in both directions</u>. In other words the <u>products</u> of the reaction can be <u>turned back</u> into the original <u>reactants</u>.

Here are some <u>examples</u> you should know about in case they spring one on you in the Exam.

The Thermal decomposition of *Ammonium Chloride*

1) When <u>ammonium chloride</u> is <u>heated</u> it splits up into <u>ammonia gas</u> and <u>HCl gas</u>.
2) When these gases <u>cool</u> they recombine to form <u>solid ammonium chloride</u>.

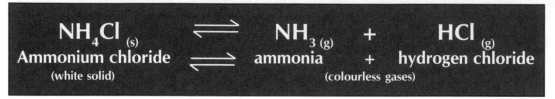

$$NH_4Cl_{(s)} \rightleftharpoons NH_{3(g)} + HCl_{(g)}$$

Ammonium chloride   ammonia   +   hydrogen chloride
(white solid)   (colourless gases)

This is a <u>typical reversible reaction</u> because the products <u>recombine</u> to form the original substance <u>very easily</u>.

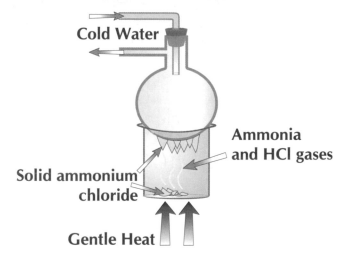

Cold Water

Ammonia and HCl gases

Solid ammonium chloride

Gentle Heat

The Thermal decomposition of *Iodine Crystals*

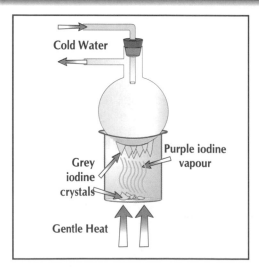

Cold Water

Grey iodine crystals

Purple iodine vapour

Gentle Heat

You can do exactly the same experiment with <u>IODINE CRYSTALS</u>, which will turn to <u>purple iodine vapour</u> and then reform as <u>grey crystals</u> when they cool.

However, do take note that this is <u>not</u> a reversible reaction, as such, but merely a <u>reversible physical process</u>.

It's just too easy

This is an easy and not especially interesting page, but I'm afraid you still need to learn it all. Make sure you understand how the apparatus in the diagrams work.

Simple Reversible Reactions

The Thermal decomposition of *hydrated copper sulphate*

1) Good old dependable <u>blue copper(II) sulphate</u> crystals here again.

2) Here they're displaying their usual trick, but under the guise of a <u>reversible reaction</u>.

3) If you <u>heat them</u> it drives the water off and leaves <u>white anhydrous</u> copper(II) sulphate powder.

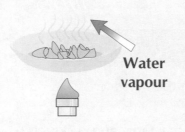

Water vapour

4) If you then <u>add</u> a couple of drops of <u>water</u> to the <u>white powder</u> you get the <u>blue crystals</u> back again.

The proper name for the <u>blue crystals</u> is <u>hydrated copper(II) sulphate</u>. "Hydrated" means "<u>with water</u>". When you drive the water off they become a white powder, <u>anhydrous copper(II) sulphate</u>. "Anhydrous" means "<u>without water</u>".

Reacting Iodine with Chlorine to get *Iodine Trichloride*

There's quite a jolly <u>reversible reaction</u> between the mucky brown liquid of <u>iodine monochloride</u> (ICl), and nasty green <u>chlorine gas</u> to form nice clean yellow crystals of <u>iodine trichloride</u> (ICl_3).

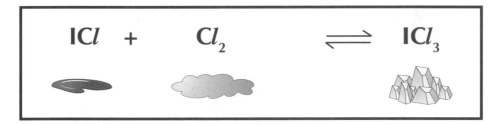

$$ ICl \ + \ Cl_2 \ \rightleftharpoons \ ICl_3 $$

1) Which way the reaction goes depends on the <u>concentration of chlorine gas</u> in the air around.

2) <u>A lot of chlorine</u> will favour formation of the <u>yellow crystals</u>.

3) <u>A lack</u> of chlorine will encourage the crystals to <u>decompose</u> back to the horrid brown liquid.

Learn these simple reactions, then see what you know

These reactions might seem a bit obscure but they're all mentioned in one syllabus or another, so any of them could come up in your Exam. There really isn't much to learn here. <u>Scribble it</u>.

Reversible Reactions in Equilibrium

A <u>reversible reaction</u> is one where the <u>products</u> can react with each other and <u>convert back</u> to the original chemicals. In other words, <u>it can go both ways</u>.

> A ***reversible reaction*** is one where the ***products*** of the
> reaction can ***themselves react*** to produce the ***original reactants***
>
> A + B $\rightleftharpoons$ C + D

*Reversible Reactions will reach **Dynamic Equilibrium***

1) If a reversible reaction takes place in a <u>closed system</u> then a state of <u>equilibrium</u> will always be reached.

2) <u>Equilibrium</u> means that the <u>relative (%) quantities</u> of reactants and products will reach a certain <u>balance</u> and stay there. "A closed system" just means that none of the reactants or products can <u>escape</u>.

Dynamic Equilibrium

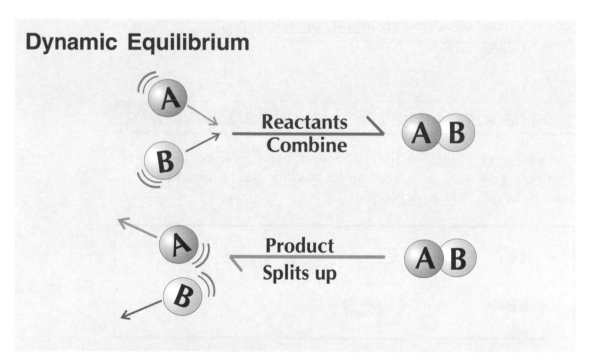

3) It is in fact a <u>DYNAMIC EQUILIBRIUM</u>, which means that <u>the reactions are still taking place</u> in <u>both directions</u> but the <u>overall effect is nil</u>. This is because the forward and reverse reactions <u>cancel each other out</u> — the reactions are taking place at <u>exactly the same rate</u> in both directions.

Dynamic Equilibrium — not half as exciting as it sounds

The idea of Dynamic Equilibrium is a bit like trying to walk up a down escalator. Both you and the escalator are moving, but because you're both moving in opposite directions at the same rate, the upshot is that you go nowhere. Perhaps Dynamic Equilibrium is also a metaphor for life...

Reversible Reactions in Equilibrium

Changing Temperature and Pressure to get *More Product*

1) In a reversible reaction the "position of equilibrium" (the relative amounts of reactants and products) depends very strongly on the temperature and pressure surrounding the reaction.

2) If we deliberately alter the temperature and pressure we can move the "position of equilibrium" to give more product and less reactants.

Two *very simple rules* for which way the equilibrium will move

1) **All reactions are exothermic in one direction and endothermic in the other.**

If we raise the temperature, the endothermic reaction will increase to use up the extra heat.
If we reduce the temperature, the exothermic reaction will increase to give out more heat.

2) **Many reactions have a greater volume on one side, either of products or reactants.**

If we raise the pressure it will encourage the reaction which produces less volume.
If we lower the pressure it will encourage the reaction which produces more volume.

This is all summed up very nicely by **Le Chatelier's Principle**, which states:

IF YOU CHANGE THE CONDITIONS, THE POSITION OF EQUILIBRIUM WILL SHIFT TO OPPOSE THE CHANGE

(My guess is that Le Chatelier was French. Ask "Teach" whether you should know about him and his principle.)

Learning/forgetting — the worst reversible reaction of them all...

There's three sections here: the definition of a reversible reaction, the notion of dynamic equilibrium, and Le Chatelier's Principle. Make sure you know them. Examiners like asking you about what will happen if the conditions of a dynamic equilibrium are changed — make sure you know.

The Haber Process Again

Other details of the Haber Process are given on page 53.

The **Haber Process** is a controlled **Reversible Reaction**

The Equation is:

$$N_{2(g)} + 3H_{2(g)} \rightleftharpoons 2NH_{3(g)}$$

ΔH is -ve (see page 179), i.e. the <u>forward</u> reaction is <u>exothermic</u>.

Higher Pressure will Favour the **Forward** Reaction so **build it strong**...

1) On the <u>left side</u> of the equation there are <u>four moles</u> of gas ($N_2 + 3H_2$), whilst on the <u>right side</u> there are just <u>two moles</u> (of NH_3).

2) So any <u>increase in pressure</u> will favour the <u>forward reaction</u> to produce more <u>ammonia</u>. Hence the decision on pressure is <u>simple</u>. It's just set <u>as high as possible</u> to give the <u>best % yield</u> without making the plant <u>too expensive to build</u>. 200 to 350 atmospheres are typical pressures used.

(Applying Le Chatelier's principle gives the same result. The equilibrium moves in favour of more ammonia because that will reduce the volume of gas in the system — i.e. if we increase the pressure, the system will try to decrease it.

Lower Temperature WOULD favour the forward Reaction **BUT**...

The reaction is <u>exothermic</u> in the forward direction, which means that <u>increasing</u> the temperature will actually move the equilibrium <u>the wrong way</u>, away from ammonia and more towards H_2 and N_2.

<u>But they increase the temperature anyway</u>... this is the tricky bit so learn it really well:

Learn this really well:

1) The <u>proportion</u> of ammonia at equilibrium can only be increased by <u>lowering</u> the temperature.

2) But instead they <u>raise</u> the temperature and accept a <u>reduced proportion</u> (or <u>yield</u>) of ammonia.

3) The reason is that the <u>higher</u> temperature gives a <u>much higher RATE OF REACTION</u>.

4) It's better to wait just <u>20 seconds</u> for a <u>10% yield</u> than to have to wait <u>60 seconds</u> for a <u>20% yield</u>.

5) Remember, the unused hydrogen, H_2, and nitrogen, N_2, are <u>recycled</u> so <u>nothing is wasted</u>.

Learn the conditions and why they're the best compromise

The Haber Process is an Exam classic. Make sure you know it like your own back garden. If you don't get an Exam question on this, I'll eat a copy of this rather weighty book. LEARN IT.

The Haber Process Again

The Iron Catalyst *Speeds up* the reaction and keeps *costs down*

1) The <u>iron catalyst</u> makes the reaction go <u>quicker</u>, which gets it to the <u>equilibrium proportions</u> more quickly. But remember, the catalyst <u>doesn't</u> affect the <u>position</u> of equilibrium (i.e. the % yield).

2) <u>Without the catalyst</u> the temperature would have to be <u>raised even further</u> to get a <u>quick enough</u> reaction and that would <u>reduce the % yield</u> even further. So the catalyst is very important.

3) <u>Removing product</u> would be an effective way to improve yield because the reaction keeps <u>chasing equilibrium</u> while the product keeps <u>disappearing</u>. Eventually <u>the whole lot</u> would be converted.

4) This <u>can't be done</u> in the Haber Process because the ammonia can't be removed until <u>afterwards</u> when the mixture is <u>cooled</u> to <u>condense out</u> the ammonia.

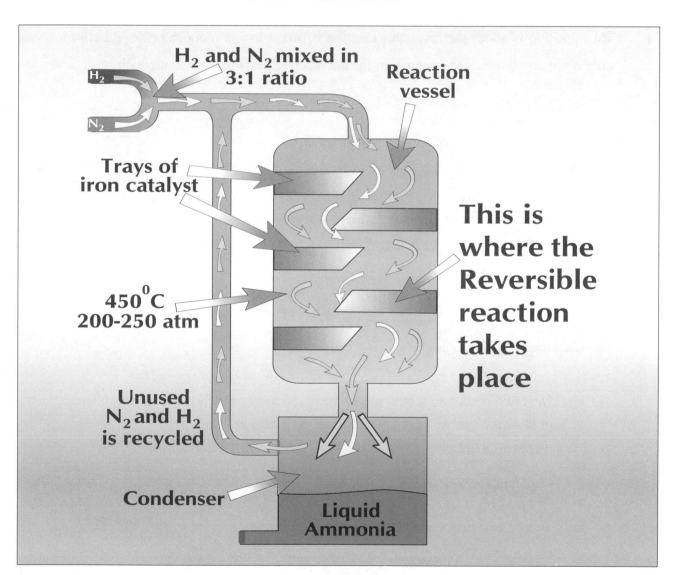

Learning the Haber Process — it's all ebb and flow

The trickiest bit is that the temperature is raised not for a better equilibrium, but for speed. Try the mini-essay method to <u>scribble down all you know</u> about equilibrium and the Haber Process.

Warm-Up and Worked Exam Questions

The basics of reversible reactions are fairly simple. However, the details of dynamic equilibrium are a bit harder — have a go at these questions to make sure you've got it.

Warm-up Questions

1) Ammonia reacts with hydrogen chloride to make ammonium chloride. Is this a reversible reaction?

2) What happens to blue hydrated copper(II) sulphate crystals when they are heated?

3) What is a closed system?

4) If a reaction is endothermic in the forward direction, can it be endothermic in the backward direction as well?

Worked Exam Question

This worked example covers the basics of reversible reactions quite nicely.

1   This question is about the reversible reaction between iodine monochloride and chlorine gas.

a)   Write down a balanced equation for the reaction between iodine monochloride and chlorine.

$ICl + Cl_2 \rightleftharpoons ICl_3$ *You must use the funny reversible reaction arrow, not the ordinary forward arrow.*

(2 marks)

b)   Iodine monochloride is a brown liquid. Iodine trichloride is a yellow crystalline solid.

i)   Describe what you would see in the reaction vessel when a high concentration of chlorine gas was present. Explain your answer.

You would see a large number of yellow crystals.

A high concentration of Cl_2 favours the forward reaction

When there's plenty of the   *and the formation of ICl_3*
reactants from the LHS of the
equation, the reaction goes
forwards.

(2 marks)

ii)   Describe what you would see in the reaction vessel when a very low concentration of chlorine gas was present. Explain your answer.

You would see relatively few yellow crystals.

Low concentration of Cl_2 favours formation of   *If there's not much of the reactants from the LHS of the equation, the reaction goes backwards.*

ICl.

(2 marks)

c)   Give an example of another reversible reaction.

Thermal decomposition of ammonium chloride

You could also have hydration/dehydration   *(2 marks)*
of copper (II) sulphate crystals.

Exam Questions

1   This question is about reversible reactions and equilibrium.

　　a)   A reversible reaction in a closed system reaches dynamic equilibrium.
　　　　 Describe what is meant by dynamic equilibrium.

　　　　 ....

　　　　 ....
　　　　　　　　　　　　　　　　　　　　　　　　　　　　　　　　　　　　　　　(2 marks)

　　b)   The reaction between reactant A and reactant B to produce product AB is reversible.

　　　　 i)   The reaction between reactant A and reactant B to produce product AB is
　　　　　　　exothermic. What kind of reaction is the decomposition of AB into products A and B?

　　　　　　　 ....
　　　　　　　　　　　　　　　　　　　　　　　　　　　　　　　　　　　　　　(1 mark)

　　　　 ii)   If the temperature in the reaction vessel is raised, which reaction will speed up,
　　　　　　　and which reaction will slow down?

　　　　　　　 ....

　　　　　　　 ....
　　　　　　　　　　　　　　　　　　　　　　　　　　　　　　　　　　　　　　(2 marks)

　　　　 iii) If the temperature is raised, how will the concentration of AB change?

　　　　　　　 ....

　　　　　　　 ....
　　　　　　　　　　　　　　　　　　　　　　　　　　　　　　　　　　　　　　(2 marks)

　　c)   State Le Chatelier's Principle.

　　　　 ....

　　　　 ....

　　　　 ....
　　　　　　　　　　　　　　　　　　　　　　　　　　　　　　　　　　　　　　(2 marks)

2   a)   Give a definition of a reversible reaction.

　　　　 ....

　　　　 ....
　　　　　　　　　　　　　　　　　　　　　　　　　　　　　　　　　　　　　　(2 marks)

　　b)   When a reversible reaction is at equilibrium, the concentrations of reactants and products
　　　　 stay the same. Has the reaction stopped? Explain your answer.

　　　　 ....

　　　　 ....
　　　　　　　　　　　　　　　　　　　　　　　　　　　　　　　　　　　　　　(2 marks)

Exam Questions

3   Ammonia is made using the Haber Process. This question is about the industrial
    conditions chosen for the Haber Process.

a)   The Haber Process involves a reversible reaction. It takes place at 200-350
     atmospheres pressure and 450°C with an iron catalyst.
     The equation for the reaction in the Haber Process is $N_{2(g)} + 3H_{2(g)} \rightleftharpoons 2NH_{3(g)}$

   i)   Explain why the Haber Process is carried out at high pressure.

        ....

        ....

        ....
                                                                                                    (2 marks)

   ii) Lower temperatures than 450°C would give higher yields. Why is the reaction
        carried out at 450°C?

        ....

        ....

        ....
                                                                                                    (3 marks)

   iii) How would the yield be affected if ammonia was removed as the reaction took
        place? Explain your answer

        ....

        ....

        ....
                                                                                                    (3 marks)

b) i)   What effect does the iron catalyst have on the yield?

        ....
                                                                                                    (1 mark)

   ii) The iron catalyst speeds up the reaction without increasing the temperature.
        Why is this particularly useful in the Haber Process?

        ....

        ....

        ....
                                                                                                    (2 marks)

Energy Transfer in Reactions

Whenever chemical reactions occur, energy is usually transferred to or from the surroundings.

In an *Endothermic Reaction*, Heat is *TAKEN IN*

> An *ENDOTHERMIC REACTION* is one which *TAKES IN ENERGY* from the surroundings, usually in the form of *HEAT*, and usually shown by a *FALL IN TEMPERATURE*

Endothermic reactions are less common and less easy to spot.
So LEARN these three examples in case they ask for one:

1) Photosynthesis is endothermic

Energy

— it takes in energy
from the sun.

Food

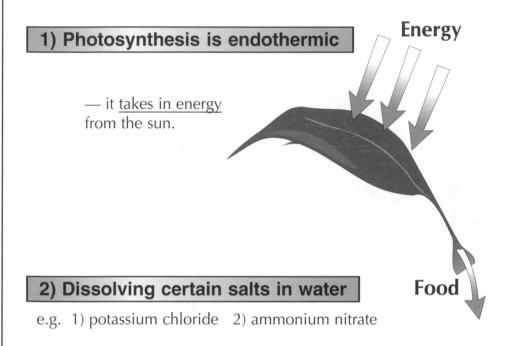

2) Dissolving certain salts in water

e.g. 1) potassium chloride   2) ammonium nitrate

3) Thermal decomposition

Heat must be supplied to cause the compound to decompose.
The best example is converting calcium carbonate into quicklime
(calcium oxide — see page 51).

$$CaCO_3 \rightarrow CaO + CO_2$$

A lot of heat energy is needed to make this happen.
In fact the calcium carbonate has to be heated in a kiln and kept at about 800°C.
It takes almost 30,000kJ of heat to make 10kg of calcium carbonate decompose.
That's pretty endothermic I'd say.

Endothermic — heat is taken in

Not too much to learn on this page. We've already come across some endothermic reactions in the book, especially thermal decompositions. Have a quick read, cover and scribble anyway.

Energy Transfer in Reactions

In an *Exothermic Reaction*, Heat is GIVEN OUT

> An *EXOTHERMIC REACTION* is one which <u>GIVES OUT</u> <u>ENERGY</u> to the surroundings, usually in the form of <u>HEAT</u> and usually shown by a <u>RISE IN TEMPERATURE</u>

1) The best example of an <u>exothermic</u> reaction is <u>burning fuels</u>. This obviously <u>gives out a lot of heat</u> — it's very exothermic.

2) <u>Neutralisation reactions</u> (acid + alkali) are also exothermic.

3) Addition of water to anhydrous <u>copper(II) sulphate</u> to turn it into blue crystals <u>produces heat</u>, so it must be <u>exothermic</u>.

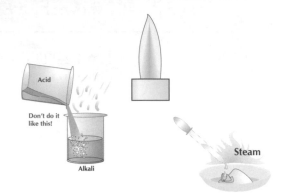

Energy is Transferred when bonds are *Formed* or *Broken*

Energy Must Always be Supplied to Break Bonds
Energy is Always Released When Bonds Form

1) During a chemical reaction, <u>old bonds are broken</u> and <u>new bonds are formed</u>.

2) Energy must be <u>supplied</u> to break <u>existing bonds</u> — so bond breaking is an <u>endothermic</u> process.

3) Energy is <u>released</u> when new bonds are <u>formed</u> — so bond formation is an <u>exothermic</u> process.

BOND BREAKING - <u>ENDOTHERMIC</u>

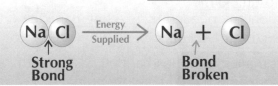

BOND FORMING - <u>EXOTHERMIC</u>

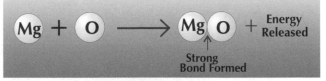

4) In an <u>exothermic</u> reaction, the energy <u>released</u> in bond formation is <u>greater</u> than the energy used in <u>breaking</u> old bonds.

5) In an <u>endothermic</u> reaction, the energy <u>required</u> to break old bonds is <u>greater</u> than the energy <u>released</u> when <u>new bonds</u> are formed.

Energy Transfer is a really important concept

You're going to need to understand this page really well before you can go on to the graphs and calculations. Cover the page, and scribble down the definitions of endothermic and exothermic.

Energy Transfer in Reactions

Energy Level Diagrams *show if it's Exo- or Endo-thermic*

In **exothermic** reactions   ΔH is -ve

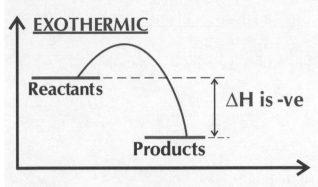

1) This shows an <u>exothermic reaction</u> because the products are at a <u>lower energy</u> than the reactants.

2) The <u>difference in height</u> represents the <u>energy given out</u> in the reaction (per mole). ΔH <u>is -ve in this case</u>.

3) The <u>initial rise</u> in the line represents the energy needed to <u>break the old bonds</u>. This is the <u>activation energy</u>.

In **endothermic** reactions   ΔH is +ve

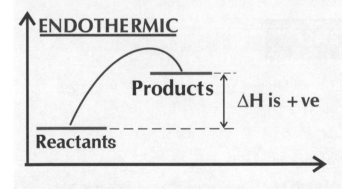

1) This shows an <u>endothermic reaction</u> because the products are at a <u>higher energy</u> than the reactants. ΔH <u>is +ve</u>.

2) The <u>difference in height</u> represents the <u>energy taken in</u> during the reaction.

The Activation Energy is **Lowered** *by* **Catalysts**

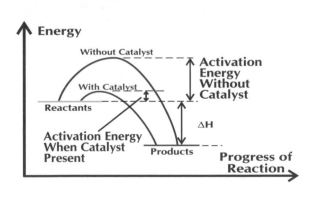

1) The <u>activation energy</u> represents the <u>minimum energy</u> needed by <u>reacting particles</u> for the reaction to occur.

2) A <u>catalyst</u> makes reactions happen <u>easier</u> (and therefore quicker) by <u>reducing</u> the initial energy needed.

3) This is represented by the <u>lower curve</u> on the diagram showing a <u>lower activation energy</u>.

4) The <u>overall energy change</u> for the reaction, ΔH, <u>remains the same</u> though.

So many graphs — make sure you know what they ALL mean

These are really important. Once you think you've learnt it, cover the page and sketch the top two graphs. Then sketch on them a line representing a catalysed reaction — it doesn't have to be exact.

Energy Transfer in Reactions

Bond Energy Calculations — need to be *practised*

1) <u>Every</u> chemical bond has a particular <u>bond energy</u> associated with it.

2) This <u>bond energy</u> is always the same no matter what <u>compound</u> the bond occurs in.

3) We can use these <u>known bond energies</u> to calculate the <u>overall energy change</u> for a reaction.

4) You need to <u>practise</u> a few of these, but the basic idea is really very simple.

Example: The Formation of HCl

The bond energies we need are: H—H   +436kJ/mol;
                                                    Cl—Cl   +242kJ/mol;
                                                    H—Cl   +431kJ/mol.

Using these known bond energies we can <u>calculate</u> the <u>energy change</u> for this reaction:

$$H_2 \ + \ Cl_2 \ \rightarrow \ 2HCl$$

1) <u>Breaking</u> one mole of H—H and one mole of Cl—Cl bonds <u>requires</u> 436 + 242 = <u>+678kJ</u>

2) <u>Forming</u> <u>two</u> moles of H—Cl bonds <u>releases</u> 2×431 = <u>862kJ</u>

3) <u>Overall</u> there is more energy <u>released</u> than used:

$$862 - 678 \ = \underline{184kJ/mol} \ \text{released.}$$

4) Since this is energy <u>released</u>, then if we wanted to show ΔH we'd need to put a <u>–ve</u> in front of it to indicate that it's an <u>exothermic</u> reaction, like this:

$$\Delta H = \underline{-184kJ/mol}$$

Energy transfers and heat — make sure you take it in

The stuff about exothermic and endothermic reactions is really quite simple. You've just got to get used to the big words. The bond energy calculations though, now they need quite a bit of practice, as do any kind of calculation questions. You can't just do one or two and think that'll be OK. No way, you've got to do loads of them. I'm sure "Teach" will provide you with plenty of practice though.

Warm-Up and Worked Exam Question

Endothermic and exothermic are simply big words for straightforward ideas. Seeing them in terms of bond energy is a bit trickier. Calculations relating to bond breaking and making need lots of attention.

Warm-up Questions

1) What's the name for a reaction which gives out heat?
2) What's the name for a reaction which takes in heat?
3) When ammonium nitrate is dissolved in water, the temperature of the water decreases. Is this an exothermic or an endothermic reaction?
4) In a chemical reaction, bonds are broken. Does bond breaking take in energy, or give out energy?

Worked Exam Questions

This worked exam question is relatively OK. You could get one just like this in the Exam. Notice how you could be asked to give an example of an endothermic reaction.

1   This question is about energy transfer in reactions.

    a) i)   What is an exothermic reaction?

An exothermic reaction is one which gives out energy to the surroundings, usually in the form of heat and shown by a rise in temperature.   *Learn the definition so you can quote it just like that.*

(1 mark)

       ii) Which of the following energy level diagrams shows an exothermic reaction?

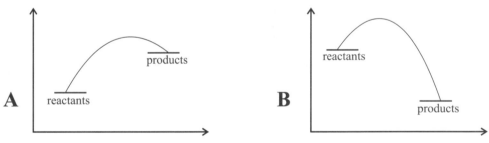

Diagram B is an exothermic reaction — products are at lower energy than reactants.

(1 mark)

    b) Give two examples of endothermic reactions.

Photosynthesis

          You could also have dissolving potassium chloride in water, or dissolving ammonium nitrate in water.

   and

Thermal decomposition of calcium carbonate

(2 marks)

Exam Questions

2   Calculate the energy change ΔH for the following reaction.
     $CH_4 + Cl_2 \rightarrow CH_3Cl + HCl$

     Bond energies required: C – H +414 kJ/mol;   C – Cl +331 kJ/mol
                                  Cl – Cl +242kJ/mol; H – Cl +431 kJ/mol

     Bonds broken = one C-H bond and one Cl-Cl bond.

     Bonds made = one C-Cl bond and one H-Cl bond.

     Bonds broken = 414 + 242 = 656 kJ/mol

     Bonds made = 331 + 431 = 762 kJ/mol

     Energy change = energy needed to break bonds - energy from making

     bonds   *Be careful to get it the*
                                                              right way around.

     Energy change ΔH = 656-762

     ΔH = -106 kJ/mol   *Remember the units.*

                                                              (4 marks)

           Negative ΔH —there's more energy released than there is used
                    — the reaction is exothermic.

Exam Questions

1   a)   Give an example of an exothermic reaction.

          ....
                                                                              (1 mark)

     b)   Look at this diagram of the energy change during a reaction.

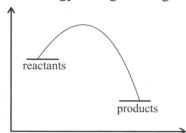

          i)   Is ΔH positive or negative in this case?

                ....
                                                                              (1 mark)

          ii)   Mark and label the activation energy on the diagram.
                                                                              (1 mark)

     c)   When a reaction takes place, old bonds are broken and new bonds are formed.
          Which process is exothermic?

          ....
                                                                              (1 mark)

Exam Questions

2   This diagram shows the energy change during a reaction. No catalyst is involved.

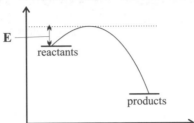

a)   What is represented by E on the diagram?

    ....
    (1 mark)

b)   Sketch a curve on the diagram to show how the reaction would proceed when a catalyst is present.

    (1 mark)

c)   How is the overall energy change different when a catalyst is involved?

    ....
    (1 mark)

3   Ethene is burnt in oxygen, producing heat. The equation for this reaction is as follows:

$$C_2H_4 + 3O_2 \rightarrow 2CO_2 + 2H_2O$$

The structure of ethene is shown on the right.

$$\underset{H}{\overset{H}{>}}C=C\underset{H}{\overset{H}{<}}$$

Bond energies required: C – H +414 kJ/mol;   C = C +615 kJ/mol
                          O = O +498 kJ/mol;   C = O +749 kJ/mol;
                          H – O +463 kJ/mol

Calculate the energy released when one mole of ethene is burnt in air.

    ....

    ....

    ....

    ....

    ....

    ....

    ....

    (5 marks)

Revision Summary

This section isn't too bad. I suppose some of the stuff on Rates of Reaction and Equilibrium gets a bit chewy in places, but the rest is all a bit of a breeze really, isn't it? Anyway, here's some more questions. Remember, if you can't answer one, look at the appropriate page and learn it. Then go back and try them again. Your hope is that one day you'll be able to glide effortlessly through all of them — it's a nice trick if you can do it.

1) What are the three different ways of measuring the speed of a reaction?

2) What are the four factors which the rate of reaction depends on?

3) Explain how each of these four factors increase the *number of collisions* between particles.

4) What is the other aspect of collision theory which determines the rate of reaction?

5) Which is the only physical factor which affects this other aspect of the collisions?

6) What happens when hydrochloric acid is added to marble chips?

7) Give details of the two possible methods for measuring the rate of this reaction.

8) Sketch a typical set of graphs for either of these methods.

9) Describe in detail how you would test the effect on the reaction rate of
      a) finer particles of solid   b) stronger concentration of acid   c) temperature

10) What happens when sodium thiosulphate is added to HCl? How is the rate of reaction measured?

11) Write down the equation for the decomposition of hydrogen peroxide.

12) What is the best way to increase the rate of this reaction?

13) What is the best way to measure the rate of this reaction? What will the graphs look like?

14) What is the definition of a catalyst? What does a catalyst do to the activation energy?

15) Name two specific industrial catalysts and give the processes they are used in.

16) What are enzymes? Where are they made? Give three examples of their use by people.

17) Sketch the graph for enzyme activity vs temperature, indicating any important temperatures.

18) What effect does freezing have on food? What happens when you thaw it out?

19) Give the word-equation for fermentation. Which organism and which enzyme are involved?

20) Explain what happens in brewing and bread-making. What is the difference between them?

21) What gives yoghurt and cheese their flavour?

22) Explain how biological detergents work.

23) In what three ways are enzymes protected in industrial processes?

24) What is a reversible reaction? Describe three simple reversible reactions involving solids.

25) Explain what is meant by dynamic equilibrium in a reversible reaction.

26) How does changing the temperature and pressure of a reaction alter the equilibrium?

27) How does this influence the choice of pressure for the Haber Process?

28) What determines the choice of operating temperature for the Haber Process?

29) What effect does the catalyst have on the reaction?

30) Give three examples of exothermic and three examples of endothermic reactions.

31) Draw energy level diagrams for these two types of reaction.

32) How do bond breaking and bond forming relate to these diagrams?

33) What are bond energies and what can you calculate from them?

EXTRA SECTION

THIS SECTION IS FOR TRIPLE AWARD CHEMISTRY ONLY

This section contains all the additional material you'll need to know if you are doing the separate Chemistry GCSE. If you're only doing the Double Award Science GCSE (as most people do), you won't need any of this, so don't revise it.

Acids are all about **Hydrated Protons**

1) When mixed with <u>water</u>, all acids release <u>hydrogen ions</u> — H⁺ (an H⁺ ion is just a proton). E.g.

$$HCl_{(g)} + water \longrightarrow H^+_{(aq)} + Cl^-_{(aq)}$$
$$H_2SO_{4\,(l)} + water \longrightarrow 2H^+_{(aq)} + SO_4^{2-}_{(aq)}$$

But HCl <u>doesn't</u> release hydrogen ions <u>until</u> it meets water — so hydrogen chloride gas isn't an acid.

2) The <u>positive</u> H⁺ ion attracts the slightly <u>negative</u> side of a water molecule. (Water is '<u>polar</u>' — the <u>oxygen</u>-side of the molecule is slightly <u>negative</u>, while the <u>hydrogen</u>-side is slightly <u>positive</u>.)

3) This proton with bits of water in tow is now written 'H⁺(aq)' and given the fancy name of '<u>hydrated proton</u>'. And it's these hydrated protons that make acids <u>acidic</u>, if you like.

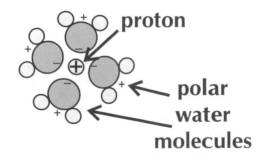

proton

polar water molecules

Now you know the truth

Yep, we've been lying to you all this time about what acids are. They're not just boring old H⁺ ions at all — they're exciting, lovely HYDRATED PROTONS. Actually, this isn't very hard at all. Once you realise that water is polar, you should have no problem remembering what they do with protons.

Acids and Bases

Bases want to Grab H+ ions

1) Not all bases <u>dissolve</u> in water, but those that do are called <u>alkalis</u>.
2) <u>Alkalis</u> form OH⁻ ions (<u>hydroxide</u> ions) when in water.

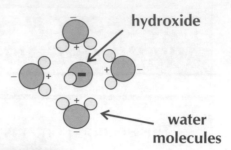

hydroxide

water molecules

Ammonia: $NH_{3(g)} + H_2O_{(l)} \longrightarrow NH_4^+{}_{(aq)} + OH^-{}_{(aq)}$

3) Hydroxide ions are also <u>hydrated</u>, i.e. they also have <u>water molecules</u> 'in tow'. But because they're <u>negatively</u> charged, it's the <u>positive</u> side of water that sticks on.
4) A scientist called <u>Arrhenius</u> reckoned that this was what made <u>bases</u> special — the fact that they <u>released</u> OH⁻ ions. (But since not all bases are <u>soluble</u>, this couldn't be the whole story.)
5) However, <u>Lowry</u> and <u>Brønsted</u> made things a bit more general. They came up with definitions that work for both <u>soluble</u> and <u>insoluble</u> bases:

<u>Acids</u> put H⁺(aq) ions into a solution — i.e. they're <u>proton donors</u>.
<u>Bases</u> remove H⁺(aq) ions from a solution — i.e. they're <u>proton acceptors</u>.

Acids can be Strong or Weak

1) <u>Strong acids</u> (e.g. sulphuric, hydrochloric and nitric) <u>ionise completely</u> in water. This means that <u>every</u> hydrogen atom is <u>released</u> to become a <u>hydrated proton</u> (so there are <u>loads</u> of H⁺(aq) ions).
2) <u>Weak acids</u> (e.g. ethanoic, citric) ionise only very <u>slightly</u>, i.e. only <u>some</u> of the hydrogen atoms in the compound are released (maybe less than 1%) — so only <u>small numbers</u> of H⁺(aq) ions are formed.

<u>Strong acid:</u> $HCl + water \longrightarrow H^+ + Cl^-$
<u>Weak acid:</u> $H_2CO_3 + water \rightleftharpoons H^+ + HCO_3^-$

Note the 'reversible reaction' symbol for a weak acid.

3) The <u>pH</u> of an acid or alkali is a measure of the <u>concentration</u> of H⁺(aq) ions in a solution. <u>Strong</u> acids have a pH of about 0 or 1, while the pH of a <u>weak</u> acid might be 4, 5 or 6.
4) The pH of an acid or alkali can be measured with a <u>pH meter</u> or with <u>Universal Indicator</u> paper (or by seeing how fast a sample reacts with, say, magnesium). See page 197 for more methods.
5) There are strong and weak <u>alkalis</u> too. The <u>hydroxides</u> of sodium and potassium, KOH and NaOH, are <u>strong</u> (pH 14), while ammonia is a <u>weak</u> alkali (pH 9-10).

'Dilute' doesn't mean it's weak

Remember, '<u>weak</u>' and '<u>strong</u>' describe the <u>proportion</u> of hydrogen atoms released, whereas '<u>dilute</u>' and '<u>concentrated</u>' just describe how '<u>watered down</u>' your acid is. An acid can be <u>dilute but strong</u>.

Solubility of Salts

Something is <u>soluble</u> if it <u>dissolves</u> — like <u>sugar</u> when you put it in tea.
Something is <u>insoluble</u> if it <u>doesn't dissolve</u> — like <u>sand</u> when you put it in tea.

Solubility — Learn the Proper **Definitions**

1) The <u>solubility</u> of a substance in a given solvent is the number of <u>grams of the solute</u> (the solid) that dissolve in <u>100 g of the solvent</u> (the liquid) at a particular <u>temperature</u>.
E.g. at <u>room temperature</u>, about 35 g of sodium chloride (NaCl) will dissolve in 100 g of water.

2) A <u>saturated solution</u> is one that cannot hold any more solid <u>at that temperature</u> — and you have to be able to see <u>solid</u> on the bottom to be certain that it's saturated.

Solubility Rules — Learn them, Okay

1) As a general rule, most <u>ionic</u> compounds are <u>soluble</u> in water, while most <u>covalent</u> compounds are <u>not</u>.

2) Unfortunately, there are quite a few rules that you need to know off by heart.

A Salt is Usually Soluble in Water if...

1) IT'S A SALT OF SODIUM, POTASSIUM OR AMMONIUM

2) IT'S A NITRATE (NO_3^-), CHLORIDE (Cl^-) OR A COMMON ETHANOATE (CH_3COO^-)
Exceptions: silver chloride, AgCl and lead chloride, $PbCl_2$

3) IT'S A SULPHATE SO_4^{2-} — but barium sulphate and lead sulphate are insoluble, and calcium sulphate is only slightly soluble.

A salt is Usually Insoluble in Water if...

4) IT'S A CARBONATE, (CO_3^{2-}) — except those of sodium, potassium and ammonium.

5) IT'S A HYDROXIDE, (OH^-) — except those of sodium, potassium and ammonium.

Solubility Curves show when a solution is **Saturated**

1) A <u>solubility curve</u> plots the <u>mass of solute</u> dissolved in a saturated solution at <u>various temperatures</u>.
2) The solubility of most solids <u>increases</u> as the temperature <u>increases</u>.
3) This means that <u>cooling</u> a saturated solution will usually cause some solid to <u>precipitate out</u>.
4) The <u>mass</u> of solute <u>precipitated</u> (or the mass of <u>crystals</u> formed) by <u>cooling</u> a solution can be calculated from a solubility curve:
Draw lines perpendicular to both axes through the temperatures in the question, then subtract the smaller mass from the larger — that difference must precipitate out on cooling.

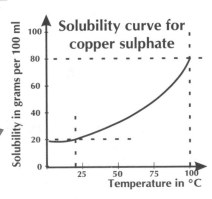

<u>Example:</u> *How much copper sulphate will precipitate out when 100 ml of a saturated solution is cooled from 100 °C to 20 °C?* <u>Answer:</u>
80 g – 20 g = <u>60 g</u>

My brain is saturated
Those five solubility rules (plus the exceptions) are worth learning properly. So cover the page and write down the types of substance that are <u>soluble</u> and insoluble together with all the <u>exceptions</u>.

Making Salts

Salts have a positive bit and a negative bit, e.g. sodium chloride (Na^+Cl^-), lead nitrate ($Pb^{2+}(NO_3^-)_2$), and so on. There are four main methods for making salts which you need to know about:

1) Reaction between the **Elements**

1) This works well when you want to combine a fairly reactive metal with a non-metal gas, e.g. reacting aluminium foil or iron wool with chlorine.

2) You just pass a reactive gas over a heated metal in a combustion tube. This method's good because you end up with a pure, dry salt.

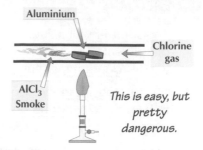

Aluminium

Chlorine gas

$AlCl_3$ Smoke

This is easy, but pretty dangerous.

E.g. $2Al + 3Cl_2 \longrightarrow 2AlCl_3$

2) Making **Insoluble** Salts — **Precipitation** Reactions

Just mix an acid and a nitrate — simple as that

1) If the salt you want to make is insoluble, you can use a precipitation reaction.

2) You just need to pick the right acid and nitrate, then mix them together. E.g. if you want to make lead chloride (which is insoluble), mix hydrochloric acid and lead nitrate.

3) Once the salt has precipitated out (and is lying at the bottom of your flask), all you have to do is filter it from the solution, wash it and then dry it on filter paper.

E.g. $Pb(NO_3)_2(aq) + 2HCl(aq) \longrightarrow PbCl_2(s) + 2HNO_3(aq)$

3) Making **Soluble** Salts — but **NOT** of **Na**, **K** or **NH₄**

1) You need to pick the right acid, plus a metal carbonate or metal hydroxide (as long as it's insoluble).

2) You add the carbonate or hydroxide to the acid until all the acid is neutralised. (The excess carbonate or hydroxide will just sink to the bottom of the flask when all the acid has reacted.)

3) Then filter out the excess carbonate, and evaporate off the water — and you should be left with a pure, dry salt.

You can't use Na, K or NH₄ carbonates or hydroxides, as they're soluble (so you can't tell whether the reaction has finished).

Filtering — to get rid of the excess carbonate or hydroxide.

E.g. you can use copper carbonate and nitric acid to make copper nitrate, or iron hydroxide and hydrochloric acid to make iron chloride.

$CuCO_3(s) + 2HNO_3 \longrightarrow Cu(NO_3)_2(aq) + CO_2 + H_2O$
$Fe(OH)_2(s) + 2HCl \longrightarrow FeCl_2(aq) + 2H_2O$

And if the metal is between calcium and copper in the reactivity series, you can just use the metal itself instead of a carbonate or hydroxide.

4) Making Salts of **Na**, **K** or **NH₄**

1) Sodium, potassium and ammonium salts are made by neutralising an acid with a hydroxide.

2) So to make sodium chloride, react hydrochloric acid with sodium hydroxide (and then evaporate off the water).

E.g. $NaOH(aq) + HCl(aq) \longrightarrow NaCl(aq) + H_2O$

Learn how to make your salts

Lots of details here. Try remembering this: IN — Insoluble salts are made with Nitrates. SHoCk — Soluble salts are made with Hydroxides and Carbonates.

Gases: Solubility and Collection

All **Gases** are **Soluble** — to some extent, anyway

1) 'Chlorine water' is a <u>solution</u> of <u>chlorine</u> gas in <u>water</u> (unsurprisingly). It's used as <u>bleach</u> for the paper and textile industries, and also to <u>sterilise</u> water supplies (as bleaching bacteria <u>kills</u> them).

2) The <u>amount</u> of gas that dissolves depends on the <u>pressure</u> of the gas above it — the <u>higher</u> the pressure, the <u>more gas</u> that dissolves. (Fizzy drinks initially contain a <u>lot</u> of dissolved carbon dioxide. But when you take the cap off, the pressure's <u>released</u> and a lot of CO_2 fizzes <u>out</u> of solution.)

3) Gases become <u>less soluble</u> as the <u>temperature</u> of the solvent <u>increases</u>, which is exactly the <u>opposite</u> of solids. (Aquatic life needs dissolved <u>oxygen</u>, but <u>pollution</u> and <u>warm water</u> discharged from towns and industry <u>raises</u> the temperature and <u>lowers</u> the dissolved oxygen levels, causing problems.)

The **Collection Method** depends on the **Gas**

A <u>side-arm flask</u> is the standard apparatus to use when you're trying to collect gases. But what you <u>connect</u> the side arm to depends on what it is you're trying to collect...

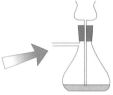

1) Gas Syringe

You can use a <u>gas syringe</u> to collect pretty much <u>any</u> gas.

2) Collection over Water

1) You can use a delivery tube to <u>bubble</u> the gas into an upside-down measuring cylinder or gas jar filled with <u>water</u>.

2) But this method's no good for collecting things like <u>hydrogen chloride</u> or <u>ammonia</u> (because they just <u>dissolve</u> in the water).

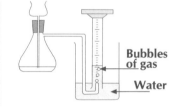

Bubbles of gas

Water

3) Upward / Downward Delivery

Use <u>upward delivery</u> to collect 'lighter than air' gases (e.g. H_2). Use <u>downward delivery</u> to collect 'heavier than air' gases (e.g. CO_2).

Upward delivery　　**Downward delivery**

Learn the **Tests** for these **Common Gases**

1) Hydrogen — burns with a squeaky pop when ignited. (This is an <u>explosive</u> gas, so be careful when you use it.)

2) Hydrogen chloride — turns damp <u>blue</u> litmus <u>red</u>. (Don't <u>breathe</u> this in — it won't do you any good.)

3) Ammonia — turns damp <u>red</u> litmus <u>blue</u>. (This has a very strong <u>smell</u>.)

4) Oxygen — relights a glowing splint.

5) Sulphur dioxide — damp orange dichromate paper goes green. (This also has a very strong <u>smell</u>.)

6) Carbon dioxide — turns limewater milky.

Squeaky pop!!

glowing splint

Limewater

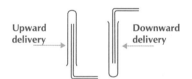
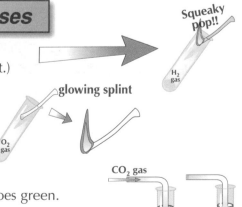

Don't let gas production escape you

On a practical note — when you're collecting a gas, it's a good idea to allow the reaction to run a little <u>before</u> connecting the actual <u>collecting apparatus</u>. This make the gas you collect purer.

Tests for Positive Ions

Got a compound, but you <u>don't know</u> what it is? The next two pages should help you identify it.

Flame Tests — Spot the Colour

Compounds of some metals burn with a characteristic colour, as you
see every November 5th. So, remember, remember…

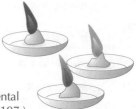

(i) <u>Sodium</u>, Na^+, burns with an orange flame.

(ii) <u>Potassium</u>, K^+, burns with a lilac flame.

(iii) <u>Calcium</u>, Ca^{2+}, burns with a brick-red flame.

(iv) <u>Copper</u>, Cu^{2+} burns with a blue-green flame.

(See also Instrumental
Methods on page 197.)

Add NaOH and look for a Coloured Precipitate

1) Many <u>metal hydroxides</u> are <u>insoluble</u> and precipitate out of solution when formed.

2) Some of these hydroxides have a <u>characteristic colour</u>.

3) The idea behind this test is to add a few drops of <u>sodium hydroxide</u> solution to 'test solutions'
of your mystery compound. Hopefully, you'll form one of these <u>insoluble hydroxides</u>.

4) If you do, the colour tells you which <u>hydroxide</u> you've got, and so what the '<u>metal bit</u>' of your
mystery compound must be…

"Metal"	Colour of precipitate	Ionic Reaction
Calcium, Ca^{2+}	White	$Ca^{2+}(aq) + 2OH^-(aq) \rightarrow Ca(OH)_2(s)$
Copper(II), Cu^{2+}	Blue	$Cu^{2+}(aq) + 2OH^-(aq) \rightarrow Cu(OH)_2(s)$
Iron(II), Fe^{2+}	Sludgy green	$Fe^{2+}(aq) + 2OH^-(aq) \rightarrow Fe(OH)_2(s)$
Iron(III), Fe^{3+}	Reddish brown	$Fe^{3+}(aq) + 3OH^-(aq) \rightarrow Fe(OH)_3(s)$
Aluminium, Al^{3+}	White at first. But then redissolves in excess NaOH to form a colourless solution.	$Al^{3+}(aq) + 3OH^-(aq) \rightarrow Al(OH)_3(s)$ then $Al(OH)_3(s) + OH^- \rightarrow Al(OH)_4^-(aq)$
Ammonium, NH_4^+	No precipitate, but the smell of ammonia (see also litmus test on next page).	

Ionic Equations show just the Useful Bits of Reactions

The reactions in the above table are <u>ionic equations</u>. Ionic equations are 'half' a full equation,
if you like. They just show the bit of the equation you're <u>interested</u> in — nothing else.

Example: $Ca^{2+}(aq) + 2OH^-(aq) \longrightarrow Ca(OH)_2(s)$

1) This shows the formation of (solid) <u>calcium hydroxide</u> from the <u>calcium ions</u> and the <u>hydroxide
ions</u> in solution. And it's the formation of this that helps identify the compound.

2) The <u>full</u> equation in the above reaction would be (if you started off with <u>calcium chloride</u>, say):
$CaCl_2(aq) + 2NaOH(aq) \longrightarrow Ca(OH)_2(s) + 2NaCl(aq)$

3) But the formation of <u>sodium chloride</u> is of no great interest here — it's not helping to <u>identify</u>
the compound, after all.

4) So the ionic equation just concentrates on the <u>good bits</u>.

Get the colours right and you're halfway there

An easy page, kind of. The principles are easy enough, but it's learning all those darn <u>colours</u> that's a
pain in the neck. Still, got to be done. Just keep covering the page and scribbling them down.

Tests for Negative Ions

It's not just positive ions you can test for. You'll be pleased to know you can also test for <u>negative ions</u>.

The Acid Test — use *Hydrochloric Acid*

The <u>gases</u> given off by salts reacting with <u>HCl</u> can be used for identification.

Carbonates (CO_3^{2-}) give off CO_2

With dilute <u>acids</u>, <u>carbonates</u> (CO_3^{2-}) give off <u>CO_2</u> (you can test for this with <u>limewater</u> — see page 189).

$$2H^+(aq) + CO_3^{2-}(s) \longrightarrow CO_2(g) + H_2O(l)$$

Sulphites (SO_3^{2-}) give off SO_2

<u>Sulphites</u>, (SO_3^{2-}), give off <u>SO_2</u> when mixed with dilute <u>hydrochloric acid</u>.
You can test for this with damp <u>dichromate paper</u> (see page 189).

$$SO_3^{2-}(s) + 2H^+(aq) \longrightarrow SO_2(g) + H_2O(l)$$

Test for *Sulphates* (SO_4^{2-}) and *Halides* (Cl^-, Br^-, I^-)

You can test for certain ions by seeing if a <u>precipitate</u> is formed after these reactions...

Sulphate ions, SO_4^{2-}

To test for a <u>sulphate</u> ion (SO_4^{2-}), add dilute HCl, followed by <u>barium chloride</u>, $BaCl_2$.
A white precipitate of <u>barium sulphate</u> means the original compound was a sulphate.

$$Ba^{2+}(aq) + SO_4^{2-}(aq) \longrightarrow BaSO_4(s)$$

Chloride, Bromide or Iodide ions, Cl^-, Br^-, I^-

To test for <u>chloride</u>, <u>bromide</u> or <u>iodide</u> ions, add dilute <u>nitric acid</u> (HNO_3), followed by <u>silver nitrate</u> ($AgNO_3$).

A <u>chloride</u> gives a <u>white</u> precipitate of <u>silver chloride</u>.

A <u>bromide</u> gives a <u>cream</u> precipitate of <u>silver bromide</u>.

An <u>iodide</u> gives a <u>yellow</u> precipitate of <u>silver iodide</u>.

$$Ag^+(aq) + Cl^-(aq) \longrightarrow AgCl(s)$$

The other reactions are the same as this, except the chloride ion is replaced with bromide or iodide.

Use *Litmus Indicator* to check for *Acidity* and *Alkalinity*

Testing for $H^+(aq)$ and $OH^-(aq)$ ions can be done using <u>red</u> or <u>blue</u> litmus indicator.
<u>Blue</u> litmus turns <u>red</u> if $H^+(aq)$ ions are present, i.e. if the solution is an <u>acid</u>.
<u>Red</u> litmus turns <u>blue</u> if $OH^-(aq)$ ions are present, i.e. if the solution is an <u>alkali</u>.

And if you add an <u>ammonium salt</u> to a <u>hydroxide</u>, ammonia's formed — it's easy to test for (see page 189).

The Test for *Nitrates* (NO_3^-) is a bit *Dangerous*

This test is a bit dangerous, so ask before you do it.

If none of the above tests tell you what the negative ion is, then test to see if it's a <u>nitrate</u>...

1) Mix some of the compound with a little <u>aluminium powder</u>.

2) Then add a few drops of <u>sodium hydroxide</u> solution and heat.

Any nitrate is reduced to <u>ammonia</u>, which can be tested for with <u>litmus</u> (see above).

$$NO_3^- \longrightarrow NH_3$$ The NaOH/Al mix is a good reducing agent and reduces NO_3 ions to NH_3.

Don't be so negative about Science — Maths is much harder...

<u>Four</u> tests — testing for <u>nine</u> different things. It's the <u>details</u> that are the trickiest things to remember, but you've just got to know them. Learn and test yourself on just one test at a time (but do them all).

Warm-Up and Worked Exam Questions

Solubility is fairly simple. The only difficult part is learning and remembering which salts are soluble. Identifying ions also depends largely on memory.

Warm-up Questions

1) What can be defined as a proton donor?
2) What can be defined as a proton acceptor?
3) Which chlorides are insoluble?
4) Which sulphates are insoluble?
5) What anion is tested for by adding hydrochloric acid and barium chloride?
6) Which anion makes a yellow precipitate with nitric acid and silver nitrate?

Worked Exam Question

This worked exam question covers solubility of salts, including solubility curves.

1   Look at this solubility graph for potassium nitrate.

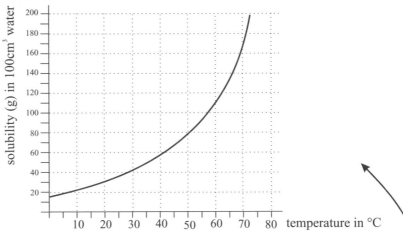

a) i)   How much potassium nitrate will dissolve in $100cm^3$ of water at 70°C?

Approximately 162g   *Read it off from the graph.*

(1 mark)

   ii)   A saturated solution of potassium nitrate is cooled from 50°C to 25°C. How much salt will crystallise out?

At 50°C 80g is dissolved, at 25°C 35g is dissolved.

80-35 = 45.   45g crystallises.

Look for the two intercepts and find the difference.   *(2 marks)*

b) Sodium hydroxide is soluble. Give an example of an insoluble hydroxide.

All common hydroxides except sodium,   _Calcium hydroxide_
potassium and ammonium are insoluble.   *(1 mark)*

c) Describe what happens when aqueous NaOH is added to a solution of copper(II) chloride.

A blue precipitate of insoluble copper hydroxide forms.

Mention the colour of any precipitate.   *(2 marks)*

Exam Questions

1   Calcium sulphate is a very slightly soluble salt. Magnesium sulphate is a soluble salt.

    a)   Calcium sulphate is produced by adding sulphuric acid to calcium nitrate solution.

        i)   Write a balanced equation for this reaction.

          ....

(2 marks)

        ii)   How do you know when the reaction is finished?

          ....

(1 mark)

    b)   Describe a suitable method of preparing magnesium sulphate.

      ....

      ....

      ....

      ....

(4 marks)

    c)   Ammonium, sodium and potassium salts are soluble. They must be prepared by titration.
        Explain why they cannot be prepared by the method you described in part (b).

      ....

      ....

      ....

(2 marks)

2   Our understanding of acids and bases has developed through history. At one time, it was
    thought that all acids ionised to form H^+ ions, while all bases ionised to form OH^- ions.

    a)   Briefly describe the ideas that Lowry and Brønsted had about acids and bases.

      ....

      ....

      ....

(2 marks)

    b)   Sulphuric acid ionises in water. Write an equation to show this.

      ....

(1 mark)

    c)   Ethanoic acid is a weak acid. Describe how the behaviour of weak acids in aqueous
        solution differs from the behaviour of strong acids in aqueous solution.

      ....

(1 mark)

194

Exam Questions

3   a) When the cap is removed from a bottle of fizzy lemonade, the lemonade fizzes up.

    i) What gas are the bubbles in the lemonade?

    ....
    (1 mark)

    ii) Explain why the bubbles appear when the cap is removed.

    ....

    ....
    (2 marks)

   b) A liquid and a solid are mixed together in a flask, producing hydrogen gas.
       Describe how you would measure the amount of gas produced.

    ....

    ....

    ....
    (3 marks)

   c) Explain why ammonia gas must not be collected over water.

    ....

    ....

    ....
    (2 marks)

4   a) Complete the following table, which shows the tests for some ions.

ion	reagent(s) added	colour of precipitate
....	NaOH	dark green
Ca^{2+}	....	....
SO_4^{2-}	....................+	white

(4 marks)

   b) An unknown solid is dissolved in water. Aqueous NaOH is added, and a white
       precipitate is formed, which dissolves in excess NaOH. Another solution of the solid is
       prepared. Dilute HNO_3 and a few drops of silver nitrate are added. A cream coloured
       precipitate forms.

       What is the formula of the unknown solid?

    ....
    (3 marks)

SEPARATE SCIENCES — ADDITIONAL MATERIAL

Quantitative Chemistry

Avogadro's Law — *One Mole* of *Gas* Occupies *24 dm³*

Learn this fact — you're going to need it:

dm³ is just another way of writing 'litre', so
1 dm³ = 1000 cm³

One mole of any gas always occupies 24 dm³ (= 24 000 cm³) at room temperature and pressure (r.t.p — 25°C and 1 atmosphere)

Example 1 What's the volume of 4.5 moles of chlorine at r.t.p.?

Answer: 1 mole = 24 dm³, so 4.5 moles = 4.5 × 24 dm³ = 108 dm³

Volume
Moles × 24

Example 2 How many moles are there in 8280 cm³ of hydrogen gas at r.t.p.?

Answer: Number of moles = $\dfrac{\text{Volume of gas}}{\text{Volume of 1 mole}}$ = $\dfrac{8.28}{24}$ = 0.345 moles

Don't forget to convert from cm³ to dm³.

Concentration = No. of Moles ÷ Volume

The concentration of a solution is measured in moles per dm³ (i.e. moles per litre).
So 1 mole of stuff dissolved in 1 dm³ of solvent has a concentration of 1 mole per dm³.

$\dfrac{M}{C \times V}$

$$\text{Concentration} = \frac{\text{Number of Moles}}{\text{Volume}}$$

Example 1 What's the concentration of a solution with 2 moles of salt in 500 cm³?

Answer: Easy — you've got the number of moles and the volume, so just stick it in the formula...

Concentration = $\dfrac{2}{0.5}$ = 4 moles per dm³

Convert the volume to litres (i.e. dm³) first by dividing by 1000.

Example 2

3 molar is sometimes written '3 M'.

How many moles of sodium chloride are in 250 cm³ of a 3 molar solution of sodium chloride?

Answer: Well, 3 molar just means it's got 3 moles per dm³. So using the formula...

Number of moles = concentration × volume = 3 × 0.25 = 0.75 moles

Crystals sometimes trap *Water of Crystallisation*

1) Magnesium sulphate crystals have a formula of $MgSO_4.6H_2O$ — this means that as the salt crystallises, six water molecules are trapped for each particle of magnesium sulphate.
2) This water evaporates when the crystals are heated, so your crystals weigh less. Cue the question...

Example 1 25 g of $CuSO_4.xH_2O$ gave 16 g of $CuSO_4$ after heating. Find x.

You have to work out how many molecules of water are trapped for each copper sulphate particle. So...

1: *Write the equation*: $CuSO_4.xH_2O(s) \longrightarrow CuSO_4(s) + xH_2O(g)$

You're told the first two, so the mass that's lost must be due to the water.

2: *Put in the masses*:   25 g $\longrightarrow$ 16   9

3: *Convert to moles*: 16 g of $CuSO_4$ is 16 / 160 = 0.1 moles.
9 g of H_2O is 9 / 18 = 0.5 moles.

For this you need the equation: No. of moles = Mass (in grams) ÷ M$_r$ (where M$_r$ is the molecular mass). It's in Section Three.

4: *Work out the ratios*: 0.1 moles of $CuSO_4$ combine with 0.5 moles of H_2O.
So 1 mole of $CuSO_4$ combines with 5 moles of H_2O.
So x must be 5 — and the formula is $CuSO_4.5H_2O$.

Concentration — makes revision so much easier

This looks worse than it is — really it's a couple of formula triangles and a 4-step method (shocker).

Titration

Titrations are used to find out the concentration of acid or alkali solutions.
They're also handy when you're making salts of soluble bases.

Titrations need to be done Accurately

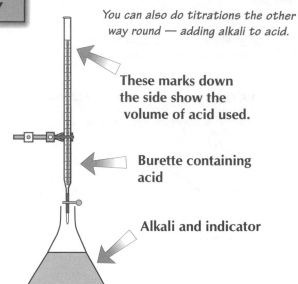

You can also do titrations the other way round — adding alkali to acid.

1) Titrations allow you to find out exactly how much acid is needed to neutralise a quantity of alkali.
2) You put some alkali in a flask, along with some indicator, e.g. phenolphthalein.
3) Add the acid, a bit at a time, to the alkali using a burette — giving the flask a regular swirl. Go especially slowly when you think the alkali's almost neutralised.
4) The indicator changes colour when all the alkali has been neutralised, e.g. phenolphthalein is pink in alkalis, but colourless in acids.
5) Record the amount of acid used to neutralise the alkali. It's best to repeat this process a few times, making sure you get the same answer each time.

These marks down the side show the volume of acid used.

Burette containing acid

Alkali and indicator

The Calculation — Work out the Numbers of Moles

Now for the calculations... basically, you're trying to find the number of moles of each substance. A formula triangle is pretty handy here. (And it's the same one as on page 195.)

Example:

Suppose you start off with 25 cm³ of sodium hydroxide in your flask, and you know that its concentration is 0.1 moles per dm³.

You then find from your titration that it takes 30 cm³ of sulphuric acid (of an unknown concentration) to neutralise the sodium hydroxide.

Find the concentration of the acid.

Step 1: *Work out how many moles of the 'known' substance you have*:
Number of moles = concentration × volume = 0.1 × (25 / 1000) = 0.0025 moles

Step 2: *Write down the equation of the reaction...*
$$2NaOH + H_2SO_4 \longrightarrow Na_2SO_4 + 2H_2O$$
...and work out how many moles of the 'unknown' stuff you must have had.

Using the equation, you can see that for every two moles of sodium hydroxide you had...
...there was just one mole of sulphuric acid.

So if you had 0.0025 moles of sodium hydroxide...
...you must have had 0.0025 ÷ 2 = 0.00125 moles of sulphuric acid.

Step 3: *Work out the concentration of the 'unknown' stuff.*
Concentration = number of moles ÷ volume
= 0.00125 ÷ (30 / 1000) = 0.0417 moles per dm³

Phenolphthalein in a flask — nothing complicated about that

The method above is the same as writing one formula triangle for the acid and one for the alkali, filling in the corners you're given, and then finding all the others. It works every time.

Instrumental Methods

Machines can be used to follow a neutralisation reaction, or more generally to analyse a substance.

You can Monitor Reactions using Machines

The neutralisation point of a reaction (i.e. when an alkali or acid is just neutralised) is usually found by using an indicator (e.g. universal indicator) in a titration, but there are other methods...

1) Use a pH Meter

This is probably the best way to monitor the pH change during a titration. You stick a probe in the solution and you get a reading on an attached machine of what the pH is.

2) Measure the Heat Change

Since a neutralisation reaction is exothermic (i.e. it gives out heat), you can measure the temperature and detect when no more heat is given off.
Even better, after the reaction's finished, excess acid will start to cool the solution.

3) Use a Conductivity Meter

1) A conductivity meter measures the conductivity of a solution (i.e. how well it conducts electricity).

2) Conductivity depends on the number of ions in solution — more ions means greater conductivity.

3) The number of ions decreases during a titration as OH^- ions are converted to H_2O by the acid's H^+ ions. After the end point, the conductivity increases again as more H^+ ions are added.

Machines can also Analyse Unknown Substances

1) Machines can also be used to analyse samples of chemicals. They're more accurate, quicker, and can detect even the tiniest amounts of substances.

2) They're useful for medical purposes, police forensic work, environmental analysis, checking whether an athlete has taken a banned substance, analysis of products in industry and so on.

3) Rapid advances in electronics and computing have made more advanced analysis easily possible.

1) Atomic Emission Spectroscopy

1) This is essentially a flame test machine, and is used for identifying elements.

2) A tiny sample is injected into a very hot flame, and the whole spectrum of light is analysed. Each element present in the sample produces a unique spectrum.

3) It's much faster and much more reliable than can be done with the human eye.

2) Infra-red Spectroscopy

1) Most carbon compounds absorb infra-red light.

2) The compound under investigation is placed in the path of infra-red radiation and the amount of each frequency that's absorbed is plotted. The pattern of absorbance is unique for every compound.
This 'fingerprint' allows identification of individual compounds.

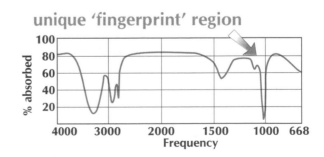
unique 'fingerprint' region

You've still got to know how to use the machine

Machines can do most of the work, but you still need to know this. For spectroscopy, make sure you remember that 'atomic emission' identifies atoms (i.e. elements), while 'infra-red' identifies compounds.

Warm-Up and Worked Exam Questions

Titration calculations can get tricky if you don't take them step by step, and concentrate on each step. The best way to do that is to practise plenty of them.

Warm-up Questions

1) What colour is phenolpthalein in alkaline solution?
2) When titrating, why do you need to swirl the flask, and go slowly near the end?
3) Atomic Emission Spectroscopy is the "grown up" version of which common lab test?
4) What kind of spectroscopy tells you what functional groups are in a molecule?

Worked Exam Questions

These worked exam questions should give you a good idea of how to work out titration problems.

1   Some sulphuric acid of unknown concentration was titrated against 20cm³ of 0.25 mol dm⁻³ sodium hydroxide, using phenolpthalein as an indicator.

   a)   What piece of equipment would usually be used to add the acid very gradually?

   A burette
   (1 mark)

   b)   Write out a balanced equation for the reaction.

   $H_2SO_4 + 2NaOH \rightarrow Na_2SO_4 + 2H_2O$
   (1 mark)

   c)   Exactly 25cm³ of sulphuric acid was required to neutralise the sodium hydroxide. Find the concentration of the sulphuric acid.

   moles NaOH = 20÷1000 × 0.25 = 0.005   1000cm³ in 1 dm³, so divide by 1000 to get number of dm³, then multiply by conc. to get number of moles.

   reacting ratio = 2 moles NaOH to 1 mole H_2SO_4

   moles acid = 0.0025   so moles of acid is half moles of alkali

   moles acid in 1dm³ = (0.0025 × 1000)÷25

   moles acid in 1dm³ = 0.1.   so...   conc = 0.1mol dm⁻³
   (4 marks)

2   The formula for anhydrous sodium tetraborate is $Na_2B_4O_7$. 19.1g of hydrated sodium tetraborate crystals are heated to drive off all the water. 10.1g of the anhydrous salt remains. What is the formula for hydrated sodium tetraborate crystals, including water of crystallisation?

   1 mole anhydrous tetraborate weighs 202g.   10.1g = 0.05 moles.

   0.05 moles of hydrated tetraborate = 19.1g   so 1 mole = 382g

   Water per mole = 382-202 = 180g

   180g = 10 moles water.   therefore... $Na_2B_4O_7.10H_2O$
   (4 marks)

Exam Questions

1   Washing soda is hydrated sodium carbonate, $Na_2CO_3.xH_2O$.

Sodium carbonate solution is alkaline.

Jonathan was asked to find the mass of sodium carbonate, Na_2CO_3, in a sample of washing soda. He then used this mass to find the formula of the hydrate. He did the experiment as follows:

1:     Jonathan dissolved 14.3g of the hydrated salt in distilled water to make 1.00 dm^3 of solution.

2:     Jonathan then measured out 50.0 cm^3 of this solution with a pipette and titrated it with some 0.2 mol dm^{-3} hydrochloric acid (HCl), in the presence of methyl orange indicator. The reaction that took place is shown by the following equation:

$$Na_2CO_3 + 2HCl \rightarrow 2NaCl + H_2O + CO_2$$

a)   Jonathan did the titration and found that 25.0 cm^3 of the hydrochloric acid was required to neutralise the Na_2CO_3. Use this figure to calculate the concentration of the sodium carbonate (Na_2CO_3) solution in mol dm^{-3} and hence the number of moles of sodium carbonate in 14.3g of the hydrate.

....

....

....

....

....

....

(4 marks)

b)   Use the previous result to find the formula of the hydrated sodium carbonate. You can use any method you like but your working must be clear.

(Relative atomic masses: H = 1.00, C = 12.0, O = 16.0, Na = 23.0)

....

....

....

....

....

....

(4 marks)

Exam Questions

2　Here is the equation for a reaction which produces potassium chloride.

$$KOH + HCl \rightarrow KCl + H_2O$$

You are provided with suitable solutions of potassium hydroxide and hydrochloric acid, an indicator and apparatus to use in a titration.

Describe how you could use this reaction to produce a pure, neutral potassium chloride solution using the equipment provided.

....

....

....

....

....

....

....

....

(5 marks)

3　Nita titrates some sulphuric acid against 50cm³ potassium hydroxide. Instead of using an indicator such as phenolpthalein or methyl orange, she measures the conductivity of the solution. She adds the sulphuric acid 0.5cm³ at a time, and measures the conductivity of the solution after each addition. She records her results in a graph.

(a)　What volume of sulphuric acid has neutralised the 50cm³ hydroxide solution?

....

(1 mark)

(b)　During the titration, the numbers and types of ions present change. Explain the shape of the graph in terms of these changes.

....

....

....

(3 marks)

Water

The Sea is **Sodium Chloride** and **Other Salts** in Solution

1) <u>Sea water</u> (also known as <u>brine</u>) is a solution of many <u>different salts</u> — mainly <u>sodium chloride</u>.

2) Salt can be extracted from the sea by <u>evaporation</u>.

3) Shallow lagoons (or "<u>salt pans</u>") are flooded with sea water. The Sun <u>evaporates the water</u> and <u>leaves the salt behind</u>.

Salt — tonnes of it, in fact.

Dissolved Carbon Dioxide Makes Water **Slightly Acidic**

1) <u>Carbon dioxide</u> is quite soluble in water.

2) It dissolves to form <u>carbonic acid</u> (H_2CO_3), a <u>weak</u> acid (see page 186 for more info), and so water tends to be slightly acidic.

$$CO_2 + H_2O \rightarrow H_2CO_3$$
$$H_2CO_3 \rightleftharpoons H^+(aq) + HCO_3^-(aq)$$

Tap Water Must be **Purified**

Fresh water contains <u>impurities</u>, which need to be <u>removed</u> before it's supplied to our houses:

   i)   <u>micro-organisms</u>,
   ii) <u>clay particles</u> (a colloid, see page 202),
   iii) chemicals causing <u>bad taste/bad smell</u>,
   iv) too much <u>acidity</u>.

1) The micro-organisms are removed by adding <u>chlorine</u> — this kills the bacteria.

2) <u>Tiny colloidal particles</u> are removed from the water by adding <u>aluminium sulphate</u>. The positive aluminium ions (Al^{3+}) bond onto negatively charged <u>clay</u> particles, causing them to <u>precipitate</u> out.

3) The water is then left to <u>settle</u>, and <u>filtered</u>.

4) <u>Nasty tastes and odours</u> are removed by passing the water through "<u>activated carbon</u>" filters, or with "<u>carbon slurry</u>".

A slurry is a suspension of solid particles in a liquid — so it's a type of colloid.

5) <u>Excess chlorine</u> is removed with <u>sulphur dioxide</u>.

$$SO_2 + 2H_2O + Cl_2 \rightarrow 4H^+ + SO_4^{2-} + 2Cl^-$$

6) A <u>limestone slurry</u> is used to make sure the water <u>isn't too acidic</u>.

$$2H^+ + CaCO_3 \rightarrow Ca^{2+} + CO_2 + H_2O$$

Loads of Cl⁻, Ca²⁺ and SO₄²⁻ ions are in the water anyway from where it's seeped through the ground. These 'extra' ones make little difference.

Iron(III) Hydroxide in Water Causes Problems

Old iron pipes sometimes allow <u>iron(III) hydroxide</u> into drinking water. This causes a few problems.

1) Vegetables <u>turn brown</u> when cooking.

2) Tea takes on an inky colour and a <u>bitter taste</u>.

3) <u>Rusty stains</u> are left on clothes after washing.

There's a lot more in water than H_2O

Not much to learn about <u>sea water</u> except how salt is extracted. You need to <u>learn all the stages of purifying tap water</u> though. Learn all the stages, cover the page and write them down.

Colloids

Colloids are Mixtures of Tiny Particles Dispersed in a Liquid

1) A colloid consists of <u>really tiny particles</u> of one kind of stuff <u>dispersed</u> in (i.e. mixed in with) another kind of stuff.

2) The particles can be bits of <u>solid</u>, droplets of <u>liquid</u> or bubbles of <u>gas</u>.

3) The <u>particles</u> are called the <u>dispersed phase</u>. It can be <u>one</u> substance, or it can be a <u>mixture</u> of substances.

4) The <u>liquid</u> that contains the particles is called the <u>continuous phase</u>. Again, it can be <u>one</u> substance, or it can be a <u>mixture</u> of substances.

5) Colloids don't separate out because the particles are <u>so small</u>, and they're often <u>electrically charged</u> (see below).

dispersed phase (the particles) continuous phase (the liquid)

Three Kinds of Colloid — Sols, Emulsions and Foams

1) A <u>sol</u> is made from <u>solid particles</u> dispersed in liquid, e.g. clay particles in water, which don't settle out when the mixture is left to stand.

2) An <u>emulsion</u> is made from <u>droplets of one liquid dispersed in another</u>, e.g. emulsion paint or mayonnaise.

3) A <u>foam</u> is made of <u>very small gas bubbles</u> dispersed in a liquid, e.g. shaving foam.

Particles in a Colloid are Sometimes Charged

1) Water molecules are <u>polar</u> — they have a slightly negative end, and a slightly positive end.

2) <u>Charged particles</u> attract <u>water molecules</u> to them.

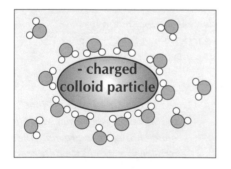  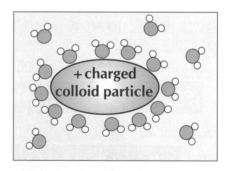

3) A <u>coating of water molecules</u> forms, keeping the particles <u>dispersed</u> in the water.

4) <u>Charged colloid particles</u> in a water-based continuous phase will <u>stay dispersed</u>, and won't clump together.

5) However, <u>metal ions</u> can cause particles of a colloid to <u>clump together</u>. They do this by attaching to the charged colloid particle and <u>neutralising the charge</u> on it.

6) When the colloid particle is <u>no longer charged</u>, it <u>no longer has a protective coating</u> of <u>water molecules</u>. So there's nothing to stop it <u>sticking to other colloid particles</u>.

Colloids are more than just mixtures

There's <u>nothing particularly difficult</u> about <u>colloids</u>. They're just tiny particles all spread out in a liquid. The <u>only slightly tricky thing</u> here is being able to <u>explain why</u> the particles don't all clump together.

Hard Water

The water in your part of the country might be <u>hard</u> or it might be <u>soft</u>.
Hardness comes from <u>limestone</u>, <u>chalk</u> and <u>gypsum</u>.

Hard Water Makes Scum and Scale

1) <u>Hard water</u> won't easily form a <u>lather</u> with soap. It makes a <u>nasty scum</u> instead.

2) Hard water also forms <u>scale</u> (calcium carbonate) on the insides of pipes, boilers and kettles. <u>Scale</u> is a bit of a <u>thermal insulator</u>. This means that a <u>kettle</u> with <u>scale on the heating element</u> takes <u>longer to boil</u> than a <u>clean</u> non-scaled-up kettle. Scale can even <u>eventually block pipes</u>. Badly scaled-up pipes and boilers need to be <u>replaced</u> — at a cost.

3) Hard water also causes a <u>horrible scum</u> to form on the <u>surface of tea</u>.

4) <u>Non-soap detergents aren't affected</u> by hard water.

Hardness is Caused by Ca²⁺ and Mg²⁺ ions

These calcium and magnesium ions come from the salts <u>calcium sulphate</u>, <u>magnesium sulphate</u>, <u>calcium carbonate</u> and <u>magnesium carbonate</u>.

1) <u>Calcium sulphate</u> and <u>magnesium sulphate</u> dissolve (just a little bit) in water.

2) <u>Calcium carbonate</u> and <u>magnesium carbonate</u> don't dissolve in water, but they will react with <u>acids</u>.

3) And since <u>CO₂</u> from the air <u>dissolves in rainwater</u> (forming <u>carbonic acid</u>, $CO_2 + H_2O \rightarrow H_2CO_3$), rainwater is slightly <u>acidic</u>.

4) This means that calcium carbonate and magnesium carbonate can react with rainwater to form <u>calcium hydrogencarbonate</u> ($H_2CO_3 + CaCO_3 \rightarrow Ca(HCO_3)_2$), which is <u>soluble</u>.

Hard Water Isn't All Bad

1) Ca²⁺ ions are good for healthy <u>teeth</u> and <u>bones</u>. They also slightly reduce the risk of <u>heart</u> disease.

2) Scale inside pipes forms a <u>protective coating</u>. It stops <u>metal ions</u>, e.g. Pb²⁺ and Cu²⁺ (from lead and copper pipes) getting into <u>drinking water</u>. It also protects iron pipes from <u>rust</u>.

Some Hardness Can be Removed

Hardness caused by dissolved <u>calcium hydrogencarbonate</u> or <u>magnesium hydrogencarbonate</u> is <u>temporary hardness</u>.
Hardness caused by dissolved <u>calcium sulphate</u> or <u>magnesium sulphate</u> is <u>permanent hardness</u>.

1) <u>Temporary hardness</u> is removed by <u>boiling</u>, e.g. calcium hydrogencarbonate <u>decomposes</u> to form insoluble CaCO₃. This <u>won't work</u> for permanent hardness, though. Heating a <u>sulphate</u> ion does <u>nothing</u>. (This calcium carbonate precipitate is the 'scale' on your kettle.)

$$Ca(HCO_3)_2(aq) \rightarrow CaCO_3(s) + H_2O(l) + CO_2(g)$$

2) <u>Both types of hardness</u> are removed by adding <u>sodium carbonate</u>. The carbonate ions join onto the calcium or magnesium ions and make an <u>insoluble precipitate</u>.

e.g. $$Ca^{2+}(aq) + CO_3^{2-}(aq) \rightarrow CaCO_3(s)$$

3) <u>Both types of hardness</u> can also be removed by '<u>ion exchange columns</u>'. These clever bits of chemistry have lots of <u>sodium ions</u> (or <u>hydrogen ions</u>) and 'exchange' them for calcium or magnesium ions.

e.g. $$Na_2Resin(s) + Ca^{2+}(aq) \rightarrow CaResin(s) + 2Na^+(aq)$$

('Resin' is a huge insoluble resin molecule.)

4) Scale is just <u>calcium carbonate</u>, and can be dissolved by <u>acid</u>.

Hard water's not that hard to learn

Hard water's <u>annoying stuff</u> — all that nasty scum and scale. To make sure you <u>really know it all</u>, take care to learn all the equations properly. You should be able to write them all down from memory.

Warm-Up and Worked Exam Questions

Water is a fairly straightforward topic, but there are some equations associated with forming and getting rid of hardness in water. As always, equations need care and thought.

Warm-up Questions

1) When carbon dioxide dissolves in water, what is formed?
2) Which cations cause hardness in water?
3) Name three nasty things that need to be removed from the domestic water supply.
4) What is a foam?
5) What is an ion exchange column for?

Worked Exam Question

This worked exam question covers a lot of what you can be asked about water hardness. Don't expect that you'll get away without having to write a balanced equation here.

1   This question is about hard water.

a)   Describe how you might compare the hardness of two samples of tap water.

You could add a drop of soap solution to each sample, and shake the sample. Keep adding soap until a lather forms which lasts 30 seconds, and record how many drops were needed. The more soap needed, the harder the water.

(3 marks)

b)   Hardness can be temporary or permanent.   *Sulphate = permanent, hydrogencarbonate = temporary*

   i)   What ions are responsible for temporary hardness in water?

   Hydrogencarbonate ions, HCO_3^-

(1 mark)

   ii)   Write a balanced equation to show temporary hardness being removed by heat.

   $Ca(HCO_3)_2 \rightarrow CaCO_{3(s)} + H_2O + CO_2$

(2 marks)

   iii)   Give two ways of removing permanent hardness from water.

   By passing the water through an ion exchange column, or by adding Na_2CO_3.   *Ion exchange column swaps calcium ions for sodium ions.*

(2 marks)

c)   Hardness in water causes pipes and heating elements to become furred with scale.

   Give an advantage of scale on lead and copper piping.

   The scale prevents poisonous copper and lead ions from dissolving in the water.

(2 marks)

Exam Questions

1   a)   What is a colloid?

....

....
(2 marks)

b)   Describe how clay particles in water are kept in colloidal suspension,
and do not coagulate.

....

....

....

....
(3 marks)

c)   Describe how aluminium sulphate removes colloidal clay from drinking water.

....

....

....

....
(3 marks)

2   Water contains impurities which must be removed before it is supplied to our homes.

a)   Why is chlorine added to the drinking water supply?

....
(1 mark)

b)   Describe how unpleasant tastes and odours are removed from the water.

....

....
(1 mark)

c)   Normal rainwater is slightly acidic.

i)   Write a balanced equation for the reaction which gives rainwater its acidity.

....
(2 marks)

ii)   How is excess acidity removed from the drinking water supply?

....

....
(1 mark)

Electrochemistry and Electrolysis

Molten Salts Conduct Electricity

A <u>salt</u> will <u>conduct an electric current</u> when <u>molten</u>. The salt is always broken up into <u>elements</u>.

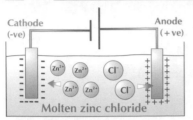

Cathode (–ve)   Anode (+ve)

Molten zinc chloride

You can melt zinc chloride using a Bunsen burner.

Electrons turn positive <u>metal cations</u> to <u>atoms</u> at the cathode (the –ve electrode):

e.g. $Zn^{2+}(l) + 2e^- \rightarrow Zn(l)$

Negative <u>anions</u> are <u>oxidised</u> (i.e. they <u>lose electrons</u>) to atoms at the <u>anode</u> (the +ve electrode): $2Cl^- \rightarrow Cl_2 + 2e^-$

Faradays and Coulombs are Amounts of Electricity

1) <u>One amp</u> flowing for <u>one second</u> means a charge of <u>one coulomb</u> has moved.

2) Generally, the amount of charge (Q, measured in coulombs) flowing through a circuit is equal to the current (I) multiplied by the time in seconds (t): <u>Q = It</u>

3) 96 000 coulombs (amps × seconds) is <u>one faraday</u>.

4) One <u>faraday</u> contains <u>one mole of electrons</u>.

> 1 A for 1 s = 1 C
> $Q = I \times t$ (seconds)
> 96 000 C = 1 faraday
> 1 faraday = 1 mole of electrons

One Mole of Product needs 'n' Moles of Electrons

A <u>sodium</u> ion needs <u>one</u> electron to make a sodium atom. So <u>one mole</u> of sodium ions is going to need <u>one mole</u> of electrons (one faraday) to make <u>one mole</u> of sodium atoms. But an ion with a <u>2+</u> charge needs <u>two moles of electrons</u> to make <u>one mole of atoms</u>, and, guess what, three for a 3+ charge...

$Na^+ + e^- \rightarrow Na$	1 mole of sodium ions + 1 mole of electrons → 1 mole of sodium atoms
$Zn^{2+} + 2e^- \rightarrow Zn$	1 mole of zinc ions + 2 moles of electrons → 1 mole of zinc atoms
$Al^{3+} + 3e^- \rightarrow Al$	1 mole of aluminium ions + 3 moles of electrons → 1 mole of aluminium atoms

Use these Steps in Example Calculations

Example: If 5 amps flows for 20 minutes during the electrolysis of aqueous lead(II) chloride ($PbCl_2$) at r.t.p., find (a) the mass of lead and (b) the volume of chlorine liberated.

1) Write out the <u>BALANCED HALF-EQUATIONS</u> for each electrode.

   $Pb^{2+} + 2e^- \rightarrow Pb$ and $2Cl^- \rightarrow Cl_2 + 2e^-$

   Writing the half-equations is easier if you remember that the full equation is: $PbCl_2 \rightarrow Pb + Cl_2$

2) Calculate the <u>NUMBER OF FARADAYS</u>.

   First calculate amps × seconds = 5 × 20 × 60 = 6000 coulombs.
   Then number of faradays = 6000 / 96 000 = 0.0625 F

3) Calculate the <u>NUMBER OF MOLES OF PRODUCT</u>.

   (divide the number of faradays by the number of electrons in the half-equations).
   0.0625 ÷ 2 = 0.03125 moles for each.

4) <u>WRITE IN THE M_r VALUES</u> from the Periodic Table to work out mass of solid products.

   Mass of lead = 207 × 0.03125 = 6.5 g

5) Use the "<u>ONE MOLE OF ANY GAS TAKES UP 24 dm³</u>" rule to find the volume of gas products.

   To find the volume of chlorine, multiply the number of moles by 24 dm³.
   Volume = 0.03125 × 24 = 0.75 dm³

More calculations I'm afraid

Electrolysis is covered in Sections Two and Three, if you need to check. What you need to learn here is how to <u>work out the amount of product</u> when you're given the <u>amount of electricity</u>.

Electrochemical Cells

You can use <u>different metals</u> and a <u>salt solution</u> to actually <u>make electricity</u>. It's how batteries work.

An Electrochemical Cell Makes Electricity

The diagram shows a <u>simple electrochemical cell</u>. It's made from <u>two different metals</u> dipped into <u>salt solutions</u> of their own ions and connected by a <u>wire</u> (not in solution).

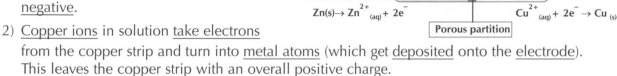

1) Atoms of the <u>more reactive metal</u> (zinc, in this case) <u>get rid</u> of <u>electrons</u> and turn into <u>positive ions</u>. The electrons are left behind on the zinc strip, making it <u>negative</u>.

2) <u>Copper ions</u> in solution <u>take electrons</u> from the copper strip and turn into <u>metal atoms</u> (which get <u>deposited</u> onto the <u>electrode</u>). This leaves the copper strip with an overall <u>positive</u> charge.

3) <u>Electrons flow through the wire</u> to the <u>less reactive metal</u> (negative to positive).

4) The <u>negative ions</u> in solution are attracted towards the <u>positive zinc ions</u>, and flow through the partition. This completes the circuit, and the <u>bulb lights up</u>.

5) Now, here's the <u>slightly weird bit</u> (pay attention):
Chemists define the <u>anode</u> (usually positive) as the electrode that <u>attracts negative ions</u>.
But in an electrochemical cell, the <u>negative ions</u> in solution flow <u>towards</u> the <u>negative</u> electrode.
This means that we call the <u>negative</u> electrode the <u>anode</u> and the <u>positive</u> electrode the <u>cathode</u>.

Confusing, I know. It's just one of those science definitions that you have to learn.

Half-Equations — What Happens at the Electrodes

This is the half-equation for what happens at the <u>anode</u>. This is <u>LOSS OF ELECTRONS — OXIDATION</u>:

$$Zn_{(s)} \rightarrow Zn^{2+}_{(aq)} + 2e^-$$

This is the half-equation for what happens at the <u>cathode</u>. This is <u>GAIN OF ELECTRONS — REDUCTION</u>:

$$Cu^{2+}_{(aq)} + 2e^- \rightarrow Cu_{(s)}$$

<u>Add them together</u> and you get the full ionic equation:

$$Zn_{(s)} + Cu^{2+}_{(aq)} \rightarrow Zn^{2+}_{(aq)} + Cu_{(s)}$$

Remember "OIL RIG"
Oxidation Is Loss (of electrons)
Reduction Is Gain (of electrons)

Remember:

OXIDATION IS LOSS OF ELECTRONS — REDUCTION IS GAIN

<u>Everything</u> that happens in an <u>electrochemical cell</u> is down to <u>loss of electrons</u> on <u>one side</u>, and <u>gain of electrons</u> on the other side. <u>That's</u> what <u>pushes the electrons</u> through the circuit. And <u>electrons moving through a circuit</u> means... ELECTRIC CURRENT.

Reduction is gain — only slightly confusing

The way to deal with this is to <u>learn carefully</u> what happens at the <u>anode</u>, and what happens at the <u>cathode</u>. It <u>doesn't matter</u> that it seems weird. All you have to do is <u>learn</u>, <u>bit by bit</u>, what's going on.

Electrochemical Cells

The **Voltage** of the Cell Depends on the **Reactivities**

The bigger the difference between the reactivity of the metal of the anode and the metal of the cathode, the more electrons move through the wire from the anode to the cathode.

That means that the bigger the difference in reactivity, the bigger the voltage of the cell.

Or, looking at it another way, a cell will shift more electrons if it's really easy for the anode to lose electrons and really easy for the cathode to gain electrons.

Standard Electrode Potentials Tell You the Voltage

1) Standard Electrode Potential sounds complicated, but all it means is "how easy it is for a metal to gain or lose electrons".

2) A list of standard electrode potentials is basically "the reactivity series with numbers". The more negative the number (or the smaller), the more reactive the element. For example, the standard electrode potential of lithium is -3.04 volts and that of silver is 0.8 volts — meaning that lithium is more reactive.

3) The potential difference or voltage of an electrochemical cell is equal to the difference in the standard electrode potentials (i.e. you subtract one number from the other to get the voltage of the cell).

4) The anode is the electrode with the most negative standard electrode potential.

Example:

What's the voltage in a cell made with nickel and lead? Nickel has a standard electrode potential of -0.25 V, and lead has a standard electrode potential of -0.13 V.

STEP 1) Write down the two numbers: -0.13 V and -0.25 V

STEP 2) Subtract the smaller number from the bigger: -0.13 − (-0.25) = +0.12 V

STEP 3) Write down which metal forms which electrode (if you need to):

Since nickel has the more negative standard electrode potential, it'll be the anode.

Fuel Cells Use **Fuel** and **Oxygen** to Make Electricity

1) A fuel cell uses the reaction between a fuel and oxygen to create a voltage. Hydrogen fuel cells are the ones you need to know about.

2) Hydrogen (the fuel) is supplied to the anode, where it gives up electrons (i.e. it's oxidised): $2H_2 \rightarrow 4H^+ + 4e^-$

3) The electrons flow through an external circuit to the cathode — this is the electric current. Also, OH^- ions move from cathode to anode to complete the circuit.

4) At the cathode, oxygen gains the electrons (i.e. it's reduced): $O_2 + 4e^- \rightarrow 2O^{2-}$

5) The overall reaction is $2H_2 + O_2 \rightarrow 2H_2O$.

6) Hydrogen fuel cells are used to provide electrical power in spacecraft such as the Space Shuttle. Some of the product of the reaction (water) is used as drinking water.

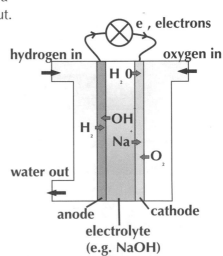

Spacecraft, fuel cells — at last, some proper science

Make sure you've definitely learnt all of the previous page. If not, go over it again. For this page, start with reactivity and voltage, then learn how to do the calculations. Finally, learn your hydrogen fuel cells.

Warm-Up and Worked Exam Questions

Electrolysis calculations take lots of little steps, and you could lose marks if you don't take good care. Electrochemical cell calculations aren't hard, but explaining what's going on in a cell can be tricky.

Warm-up Questions

1) How much charge passes through a circuit if a current of 2 amps flows for 5 minutes ?
2) How many coulombs are there in one faraday? How many electrons is that?
3) How many faradays are needed to make one mole of zinc from one mole of zinc ions?
4) In an electrochemical cell, is the more reactive metal the cathode or the anode?
5) What does OIL RIG stand for?
6) Which cell will produce the bigger voltage — one with a nickel anode and silver cathode, or one with a magnesium anode and silver cathode?

Worked Exam Question

This worked question on electrochemical cells should give you a good idea of what to write.

1 Eleanor makes a cell by dipping a plate of zinc into a solution of zinc sulphate, and a plate of copper into a solution of copper(II) sulphate. The solutions were separated by a porous partition. The two metals were connected with a wire and a bulb, in series.

(a) What happens to the bulb?

It lights up

(1 mark)

(b) Changes take place at the zinc and copper electrodes.

(i) What happens to the zinc atoms?

They lose electrons and turn into zinc ions. Zn^{2+}

(1 mark)

(ii) Write an ionic equation to show what happens at the copper electrode.

_$Cu^{2+}_{(aq)} + 2e^- \rightarrow Cu$_

(2 marks)

(c) Eleanor thinks this is a redox reaction between zinc and copper. Show that she is right.

At the copper cathode, copper ions gain electrons. They are reduced. At the zinc anode, zinc loses electrons. It is oxidised.

Remember OIL RIG:
Oxidation Is Loss, Reduction Is Gain

(3 marks)

(d) Explain how the reactions in the cell drive current through the wire.

At the zinc anode, Zn loses electrons to become Zn^{2+}.

The electrons flow around the circuit to the copper cathode.

Cu^{2+} ions gain electrons to become Cu.

(3 marks)

Worked Exam and Exam Questions

2   a)   Calculate how many moles of chlorine are made when a current of 5 amps is passed through aqueous sodium chloride solution for 30 minutes.

Charge = 5 × 30 × 60 = 9000 coulombs

Faradays = 9000 ÷ 96 000 = 0.09375

$2Cl^- \rightarrow Cl_2 + 2e^-$: so ÷ 2 to give 0.046875 moles

2 moles of electrons for each mole of chlorine.

(3 marks)

    b)   If a current of 3 amps flows through molten zinc chloride for 15 minutes, find both the mass of zinc and the volume of chlorine liberated. (Assume r.t.p.)

Charge = 3 × 15 × 60 = 2700 coulombs

Faradays = 2700 ÷ 96 000 = 0.028125

$2Cl^- \rightarrow Cl_2 + 2e^-$ and $Zn^{2+} + 2e^- \rightarrow Zn$. So ÷ 2 to give 0.01406 moles.

Mass Zn = 0.01406 × 65 = 0.91g.   Use molar mass of zinc=65 and molar volume of any gas=24dm³.

Volume Cl_2 = 0.01406 × 24dm³ = 0.34dm³

(5 marks)

1   A current is passed through aqueous sodium chloride solution using the apparatus shown.

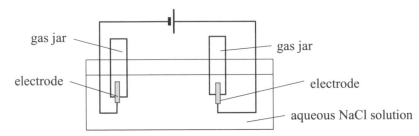

    a)   Gases are given off at the electrodes and collected.

        i)   What gas is given off at the negative electrode?

        ....

(1 mark)

        ii)   What gas is given off at the positive electrode?

        ....

(1 mark)

        iii)   After five minutes, the volume of gas in both gas jars is equal. Explain why this is.

        ....

        ....

        ....

(2 marks)

Exam Questions

2   Students make a cell as shown in the diagram.
They try using zinc, lead, magnesium and copper for the electrodes.

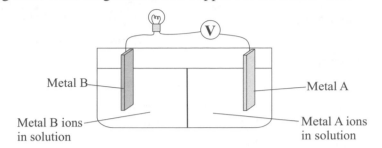

Metal B

Metal A

Metal B ions in solution

Metal A ions in solution

a)   Which two metals do you expect to give the highest reading on a voltmeter placed in the circuit? Explain your choice.

....

....

....

(3 marks)

b)   When copper and lead are used as electrodes, what would you expect to see happening around the copper electrode?

....

(1 mark)

c)   Use this information to help you calculate the voltage produced by the following cells.

Standard Electrode Potentials:

$Zn^{2+}(aq) + 2e \rightarrow Zn(s)$ -0.76 $Pb^{2+}(aq) + 2e \rightarrow Pb(s)$ -0.13

$Cu^{2+}(aq) + 2e \rightarrow Cu(s)$ 0.34 $Ag^{+}(aq) + e \rightarrow Ag(s)$ 0.80

    i)   Zinc anode, silver cathode in silver chloride solution.

....

....

(2 marks)

    ii)   Lead anode, copper cathode in copper sulphate solution.

....

....

(2 marks)

3   Hydrogen fuel cells use hydrogen, oxygen and an acidic electrolyte to make electricity.

a)   Complete this equation to show what happens at the anode.

$H_2 \rightarrow$ .... + ...

(2 marks)

b)   Complete this equation to show what happens at the cathode.

O_2 + ... + ... $\rightarrow$ ....O^{2-}

(3 marks)

Revision Summary

Try these questions without using the book. Then if you got any wrong, go back and have a look at what the answer should have been. And keep doing this until you can get all the questions right.

1) What do acids form in water?

2) How did Lowry and Brønsted define an acid? How did they define a base?

3) What makes an acid 'strong'?

4) Is a salt of sodium or potassium likely to be soluble or insoluble?

5) A saturated solution contains 50 g of solute at 60 °C and 30 g at 40 °C.
   What mass precipitates out on cooling the saturated solution from 60 °C to 40 °C?

6) Describe how you could make a soluble copper salt.

7) Name four different methods of collecting a gas.

8) Which gases must never be collected over water? Why?

9) How would you test for the gases H_2, O_2, CO_2, NH_3, HCl and SO_2?
   Describe each test, including how you would recognise a positive result.

10) What are the flame test colours of K^+, Na^+, Ca^{2+}, Cu^{2+}?

11) Describe the characteristics of the precipitates of Al^{3+}, Ca^{2+}, Cu^{2+}, Fe^{2+} and Fe^{3+} obtained after adding NaOH(aq).

12) What are the tests for CO_3^{2-} and SO_3^{2-}? What is the name of the SO_3^{2-} ion?

13) How do you test for a suspected chloride or sulphate?

14) How many moles are there in 687.5 g of $CaCO_3$? (Ca = 40, C = 12, O = 16)?

15) A salt G has a formula mass of 68. It has a hydrated salt $G.xH_2O$. On heating 7 g of the hydrated salt, 3.6 g were lost. Find x.

16) 25 cm^3 of sodium hydroxide solution, 0.15 mol dm^{-3}, was neutralised by 22.5 cm^3 nitric acid, HNO_3.
   What is the concentration of the nitric acid?

17) Name two machines that can identify unknown substances.

18) Fresh water is slightly acidic. Which gas found in the air causes this acidity, and how?

19) What ionic compound is responsible for rusty stains on laundry?

20) What is a colloid?

21) How do the charges on colloidal particles stop them from sticking together?

22) What salts produce temporary hardness? What salts produce permanent hardness?

23) Which type of hardness is removed by boiling?

24) What is an ion exchange column used for? Explain how they work.

25) If 2 amps of current flows for 3 seconds, how much charge is that, in coulombs?

26) If 3 amps flows for 30 minutes in the electrolysis of copper(II) chloride solution, find:
   a) the *mass of copper*, and b) the *volume of chlorine* formed (at r.t.p.).

27) At which electrode of a cell is a metal oxidised to a metal ion?

28) Complete the following sentence:
   The greater the difference between the reactivities of the metals that the electrodes are made from, the greater the ....

29) If a cell is made with an anode of magnesium (standard electrode potential -2.38 V) and a cathode of tin (standard electrode potential -0.14 V), what voltage is produced?

30) Write out the half-equations for a hydrogen fuel cell.

Aluminium and Titanium

Aluminium *is* Reactive, *but doesn't* Corrode

1) Aluminium is widely used (e.g. in aeroplanes, drinks cans, power cables on pylons, window frames...) because of its low density, good conductivity and corrosion resistance. But it's expensive as it's more reactive than carbon, and so it has to be extracted by electrolysis.

2) Pure aluminium is pretty soft, so it's often alloyed with other metals to make it harder and stronger.

3) A protective layer of unreactive aluminium oxide (Al_2O_3) quickly forms on the surface of aluminium. This is why aluminium doesn't corrode, despite being quite reactive.

The Oxide Layer *can be* Thickened *for Even* Greater Protection

1) It's possible to make the surface aluminium oxide layer thicker than it would naturally get — this gives even greater protection for the aluminium underneath.

2) The natural oxide layer is first removed using sodium hydroxide solution.

3) The aluminium is then made the anode of an electrolysis cell, with sulphuric acid as the electrolyte.

4) Oxygen forms on the surface of the aluminium anode, and reacts with the aluminium to form aluminium oxide. This anodising process makes the aluminium even more resistant to corrosion.

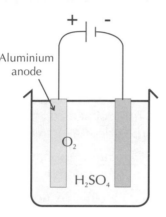

Aluminium anode

$$4OH^- \rightarrow 2H_2O + O_2 + 4e^-$$

There are some OH⁻ ions in the acid solution, even though it's an acid. The oxygen forms from these.

O_2

H_2SO_4

Impure Titanium *is not much use*

1) Titanium is a transition metal — it's better than steel for making most things — and is even more resistant to corrosion than aluminium. Also, its density is low enough for it to be used in aircraft.

2) Its big disadvantage is the cost of extracting it from its ore. Although in theory it's low enough in the reactivity series to be extracted using carbon, carbon impurities in titanium metal make it useless (unlike iron, where a little carbon makes the metal more useful, as it forms steel).

3) In fact, titanium is extracted in an inert atmosphere of argon using a more reactive metal (which first has to be obtained by the expensive process of electrolysis). So titanium turns out to be a very expensive metal.

4) This means it's only used for things where a high cost is justified, such as nuclear reactors, jet engines and replacement hip joints.

Titanium *is* Extracted *using* Sodium *or* Magnesium

1) The main ore of titanium is rutile, TiO_2.

2) Rutile is first converted to titanium chloride, $TiCl_4$.

3) This titanium chloride is then reduced using sodium or magnesium in an argon atmosphere (to prevent contamination), to leave titanium metal.

$$TiCl_4 + 4Na \rightarrow Ti + 4NaCl$$

Lots of details to learn

Remembering what aluminium and titanium are used for shouldn't be too hard, but you need to learn the details of thickening the aluminium oxide layer, and extracting titanium too. Learn, cover, scribble.

Iron, Steel and Alloys

Cast iron from a blast furnace has only limited uses. Although it's very hard and pretty rust-proof, it's also very brittle. But it's good for making products like drain covers.

Steel is an Alloy of Iron and Carbon

1) Steel is a general name given to alloys of iron that contain between 0.05% and 1.5% carbon (and often other metals too).

2) Steels are much more useful than cast iron, but they rust easily.

3) However, there are various ways to prevent steel from rusting so easily (see page 215).

Blast in Oxygen — then get rid of Oxide Impurities

1) The iron produced in a blast furnace is very impure. It contains various impurities such as carbon, manganese, silicon, phosphorus and sulphur.

2) In the first stage of producing steel, impure molten iron from the blast furnace is mixed with scrap iron (this recycling is good for profits and the environment) and, as it is heated, oxygen is blasted through.

3) This converts all the non-metal impurities to oxides. Carbon dioxide (CO_2) and sulphur dioxide (SO_2) come off as gases.

4) Calcium carbonate ($CaCO_3$) is also added — this decomposes to form calcium oxide (CaO) and carbon dioxide.

5) As calcium oxide is basic (it's a metal oxide, after all), it reacts with any non-metal (acidic) oxides, in particular silicon dioxide (SiO_2), to form 'slag'.

6) Calculated amounts of carbon and other metals are then added to give the required alloys.

	Composition	Properties	Uses
Cast iron	4% Carbon	Hard but brittle	Castings of drains, stoves, etc.
High-carbon steel	1.5% Carbon	Very hard	Cutting tools for industry, drill, etc.
Mild steel	0.25% Carbon	Quite soft, shaped 'easily' by pressing	Car bodies, tin cans, girders, etc.
Stainless steels	Contain chromium and/or nickel	Corrosion-resistant and strong	Food vessels, marine and chemical use
Titanium steel	Contains titanium	Very strong	Armour plating, for example
Manganese steel	Contains manganese	Strong, tough, and wear-resistant	Grinding machinery, caterpillar tracks

Brass and Solder are also Alloys

1) Brass is an alloy, usually of copper and zinc. Most of the properties of brass are just a mixture of those of the copper and zinc, although brass is harder than either of them.

2) Solder is an alloy of lead and tin. Unlike pure materials it doesn't have a definite melting point, but gradually solidifies as it cools down. This is pretty useful if you want to solder things together.

Alloys are really important in industry

Alloys mean a product can be made that has just the properties you want. As for the page, well, it's mostly okay, but getting your head round why using calcium carbonate is a good idea is a little tricky.

Protecting Iron and Steel

If you don't take care of steel, it'll <u>rust away</u>. Learn this page and you need never find this a problem.

There are loads of ways to **Prevent Rusting**

<u>Iron</u> and some <u>steels</u> will rust if they come into contact with <u>air</u> and <u>water</u>. To prevent this, there are various things you can do:

1) <u>Galvanise</u> it (i.e. coat it with a thin layer of <u>zinc</u>, either by dipping the iron into <u>molten</u> zinc or by <u>electrolysis</u>),

2) <u>Paint</u> it,

3) <u>Oil</u> or <u>grease</u> it,

4) Use <u>sacrificial protection</u> (i.e. attach a block of a <u>more reactive</u> metal to the iron, which corrodes <u>instead</u> of the iron).

Or use a suitable <u>alloy</u> — like stainless steel, for example.

Electroplating means **Coating** one **Metal** with **Another**

<u>Electroplating</u> means using <u>electrolysis</u> to coat one metal with a <u>layer</u> of another.

Say you want to coat an iron teapot with a thin layer of copper, then what you do is this:

1) Use the <u>teapot</u> as the <u>negative</u> electrode (<u>cathode</u>) in an <u>electrolysis</u> cell.

2) Make the <u>anode</u> out of the metal you want to form the <u>coating</u> (in this case, copper).

3) As the <u>electrolyte</u>, you need a solution containing <u>copper ions</u> (or whatever metal you want to form the coating), so in this case you could use <u>copper sulphate</u>.

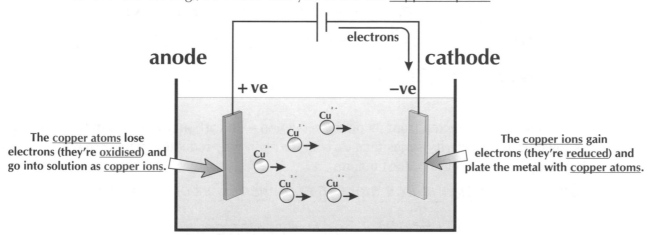

The <u>copper atoms</u> lose electrons (they're <u>oxidised</u>) and go into solution as <u>copper ions</u>.

The <u>copper ions</u> gain electrons (they're <u>reduced</u>) and plate the metal with <u>copper atoms</u>.

1) At the <u>anode</u>, electrons are <u>removed</u> from copper atoms, which then go into <u>solution</u> as ions: $Cu(s) \rightarrow Cu^{2+}(aq) + 2e^-$

2) The <u>electrons</u> flow to the teapot (the <u>cathode</u>), attracting $Cu^{2+}(aq)$ ions. These ions accept the electrons to form <u>copper atoms</u>, which stick to the teapot: $Cu^{2+}(aq) + 2e^- \rightarrow Cu(s)$

3) This all means that the <u>concentration</u> of copper in solution remains <u>constant</u>, but a very thin <u>layer</u> of copper deposits on the teapot, and an <u>identical</u> mass is lost from the copper anode.

4) Gold, silver and nickel plating are done in the same way, but the <u>cations</u> (the positive ions) in solution and the <u>anode</u> are gold, silver or nickel (obviously).

Plating is easier to learn than it looks

The half equations for <u>copper plating</u> are just the same as the ones for <u>purifying</u> copper by <u>electrolysis</u>. The only difference is whether you start with a <u>cathode</u> of <u>pure copper</u> or a cathode of <u>another metal</u>.

Warm-Up and Worked Exam Questions

It's likely that you'll get some kind of question on aluminium and titanium. The details of extracting titanium are a little tricky, so take care not to slip up there. Alloys and protecting steel are fairly easy.

Warm-up Questions

1) Why is aluminium usually alloyed?
2) What is titanium used for?
3) What's rutile?
4) Is pure iron useful for much?
5) How much carbon is there in mild steel?
6) What's high carbon steel used for?

Worked Exam Question

Take a look at this worked exam question. You should get a nice idea of what to write if you get asked a question about extracting titanium in the Exam.

1   This question is about the extraction of iron and titanium from their ores.

a)   Carbon reduces iron ore to iron. Write down a balanced equation to show this.

$$3C + 2Fe_2O_3 \rightarrow 4Fe + 3CO_2$$

— Remember it's iron(III) oxide, Fe_2O_3.

(2 marks)

b)   Can aluminium ore be reduced by carbon? Give a reason for your answer.

No, because aluminium is higher in the reactivity series than carbon   Carbon only displaces less reactive metals from their oxides.

(2 marks)

c)   Titanium is not extracted from its ore by reduction with carbon, even though titanium is lower in the reactivity series than carbon. Explain why reduction with carbon is acceptable for iron but not for titanium.

Carbon impurities in iron make it stronger. Carbon impurities in titanium make it useless.   Reducing with carbon will always leave some carbon in the metal.

(2 marks)

d)   Give brief details of the process by which titanium is extracted from titanium ore.

Titanium ore, TiO_2, is converted to titanium chloride, $TiCl_4$. The titanium chloride is reduced by sodium or magnesium metal, in a totally inert argon atmosphere.   The sodium or magnesium metal is obtained by electrolysis.

(4 marks)

SEPARATE SCIENCES — ADDITIONAL MATERIAL

Exam Questions

1   Aluminium is relatively high in the reactivity series. Reactive metals usually corrode easily, as they react readily with acids, water and oxygen.

    a)   Aluminium does not corrode easily. Explain why this is.

    ....

    ....

                                                                        (2 marks)

    b)   Aluminium's natural resistance to corrosion can be improved by anodising.

        i)   The anodising process is carried out by electrolysis. The piece of aluminium is the anode in the circuit. What chemical is used as the electrolyte?

        ....

                                                                                    (1 mark)

        ii)   Complete this ionic equation to show what happens at the aluminium anode.

               ....$OH^- \rightarrow 2$............ + ... + ...e^-

                                                                                              (3 marks)

2   Titanium is used in the manufacture of jet engines and nuclear reactors.

    a)   Give three properties of titanium that make it very useful for manufacturing.

    ....

                                                                            (3 marks)

    b)   Titanium is extracted by reducing titanium chloride with sodium metal in an inert argon atmosphere. Why does this limit the potential uses of titanium in industry?

    ....

    ....

                                                                            (2 marks)

3   This question is about steel.

    a)   Iron straight from the blast furnace contains 4% carbon. Most steels have between 0.05% and 1.5% carbon. How is the excess carbon removed from the iron to make steel?

    ....

    ....

                                                                            (1 mark)

    b)   During the production of steel, other metals can be added to the molten iron to form an alloy. Which two metals are alloyed with iron to make stainless steel?

    .... and/or ....

                                                                            (2 marks)

    c)   Steel can be galvanised to protect it from rust. What is galvanising?

    ....

                                                                             (1 mark)

You don't need to learn this page if you're doing the OCR syllabus.

Sulphuric Acid

Loads of modern industries use tonnes of <u>sulphuric acid</u>. You need to know how it's made.

The **Contact Process** is used to make **Sulphuric Acid**

1) The first stage of the <u>Contact Process</u> involves forming <u>sulphur dioxide</u> (SO_2). This is usually done by burning <u>sulphur</u> in <u>air</u> or roasting sulphide ores. (There are various <u>sulphur deposits</u> around the world, although some SO_2 comes from <u>metal extraction</u> industries — see page 214.)

$$S + O_2 \longrightarrow SO_2$$

2) The sulphur dioxide is then <u>oxidised</u> (with the help of a catalyst) to form <u>sulphur trioxide</u> (SO_3).

$$2SO_2 + O_2 \rightleftharpoons 2SO_3$$

3) Next, the sulphur trioxide is <u>dissolved</u> in concentrated sulphuric acid to form <u>fuming sulphuric acid</u>, or <u>oleum</u>.

 Dissolving SO_3 in <u>water</u> doesn't work — the reaction gives out enough <u>heat</u> to <u>evaporate</u> the sulphuric acid.

$$SO_3 + H_2SO_4 \longrightarrow H_2S_2O_7$$

4) Finally, oleum is <u>diluted</u> with measured amounts of <u>water</u> to form <u>concentrated sulphuric acid</u>.

$$H_2S_2O_7 + H_2O \longrightarrow 2H_2SO_4$$

A **Catalyst** is **Important** when making SO_3

1) The reaction in step 2 above (oxidising sulphur dioxide to sulphur trioxide) is <u>exothermic</u> (i.e. it <u>gives out</u> heat). Also, there are <u>two moles</u> of <u>product</u>, compared to <u>three moles</u> of <u>reactants</u>.

2) So <u>Le Chatelier's</u> principle (see Section Six) says that to get more <u>product</u>, you should <u>reduce</u> the <u>temperature</u> and <u>increase</u> the <u>pressure</u>.

3) Unfortunately, reducing the temperature <u>slows</u> the reaction right down, and increasing the pressure soon <u>liquefies</u> the SO_2 — no use at all.

4) With a <u>fairly high temperature</u>, a <u>low pressure</u> and a <u>vanadium pentoxide</u> catalyst, the reaction goes <u>pretty quickly</u> and you get a <u>good yield</u> (about 99%).

Conditions for Contact Process
1) Temperature: <u>450 °C</u>.
2) Pressure: <u>1-2 atmospheres</u>.
3) Catalyst: Often <u>vanadium pentoxide</u>, V_2O_5 (but others can also be used).

Sulphuric Acid can be used as a **Dehydrating Agent**

1) Sulphuric acid is used in <u>car batteries</u> (and is concentrated enough to cause severe <u>burns</u>).

2) It's also used in many <u>manufacturing</u> processes, such as making <u>fertilisers</u> and <u>detergents</u>.

$$C_6H_{12}O_6 \longrightarrow 6C + 6H_2O$$
glucose (sugar) $\longrightarrow$ carbon + water

3) And sulphuric acid is also a powerful <u>dehydrating agent</u>. It removes <u>hydrogen</u> and <u>oxygen</u> from <u>organic</u> material (like sugar), and things like hydrated <u>copper(II) sulphate</u> (which goes from blue to white as the <u>water of crystallisation</u> is removed).

$$CuSO_4.5H_2O \longrightarrow CuSO_4 + 5H_2O$$
hydrated copper(II) sulphate $\longrightarrow$ anhydrous copper(II) sulphate + water

Examiners love to ask about industrial processes

You need to understand why the <u>catalyst</u> is so vital in the <u>Contact Process</u>. Le Chatelier's principle would make it very difficult to produce sulphur trioxide — the catalyst provides a way round this.

Free Radicals

Free Radicals are Made by Breaking Covalent Bonds

1) A covalent bond, remember, is one where two atoms share electrons between them, like in H_2.

2) A covalent bond can break unevenly to form two ions, e.g. $H–H \rightarrow H^+ + H^-$.
   The H^- has both of the shared electrons, and the poor old H^+ has neither of them.

3) But a covalent bond can also break evenly — and then each atom gets one of the shared electrons,
   e.g. $H–H \rightarrow H· + H·$ The H· is called a free radical. (A free radical is shown by a dot.)

4) The free radical has one electron out of a pair, and so it's very, very reactive.

Chlorofluorocarbons contain Chlorine and Fluorine

1) Chlorofluorocarbons (CFCs for short) are organic molecules containing chlorine and fluorine,
   e.g. dichlorodifluoromethane CCl_2F_2, which is like methane but with two chlorine and two
   fluorine atoms (and an extremely long name) instead of the four hydrogen atoms.

2) Chlorofluorocarbons were used as coolants in refrigerators and air-conditioning systems.

3) CFCs were also used as propellants in aerosol spray cans.

Free Radicals from CFCs Damage the Ozone Layer

1) Ultraviolet light makes CFCs break up to form free radicals: $CCl_2F_2 \rightarrow CClF_2· + Cl·$

2) This happens high up in the atmosphere, where the ultraviolet light from the Sun is stronger.

3) Chlorine free radicals react with ozone (O_3), turning it into ordinary oxygen molecules (O_2):

$$O_3 + Cl· \rightarrow ClO + O_2$$

4) The chlorine oxide molecule ClO is very reactive, and reacts with any
   oxygen atoms around to make an oxygen molecule and another Cl· free radical.

$$ClO + O \rightarrow O_2 + Cl·$$

5) Because a chlorine free radical gets regenerated each time,
   one free radical can go around breaking up a lot of ozone molecules.

The Ozone Layer Shields the Earth from UV light

Because the ozone layer has been damaged by chlorine free radicals, more ultraviolet light
reaches the Earth's surface. Ultraviolet light is harmful to living things — in people it causes:

   i) Sunburn,   ii) Premature aging of the skin,   iii) Skin cancer.

CFCs are Banned in Some Countries

In many countries of the world CFC manufacture is banned. Butane is used as a propellant in
aerosols instead of CFCs, and old fridges that contain CFCs have to be specially dealt with so that
CFCs don't escape.

But getting agreement between all the countries of the world to ban the manufacture and use of
harmful materials is difficult.

Free radicals are really bad news
Learn the subsections one by one. First, learn what free radicals and CFCs are. Then learn what they
do to the ozone layer, and then the three ways that UV light damages living things. Simple.

Warm-Up and Worked Exam Questions

The Contact Process is a good excuse to throw in Le Chatelier's Principle, so make sure you "get" it.

Warm-up Questions

1) What forms when sulphur is burnt in oxygen?
2) Write an equation for the reaction between sulphur dioxide and oxygen.
3) What are the industrial conditions of the Contact Process?
4) Under what conditions do CFC molecules break up to form free radicals?
5) How does a hole in the ozone layer potentially damage life on the earth's surface?

Worked Exam Questions

These worked exam questions cover most of the things that you could be asked about sulphuric acid, the Contact Process and free radicals.

1   Sulphuric acid is made industrially by the Contact Process. Part of the process involves this reaction: $\quad 2SO_{2(g)} + O_{2(g)} \rightleftharpoons 2SO_{3(g)}$

(a) What catalyst is used in this reaction?

Vanadium pentoxide, V_2O_5

(1 mark)

(b) What pressure conditions would favour conversion into SO_3?   *Le Chatelier's Principle — less volume on the products' side.*

High pressure

(1 mark)

(c) In the Contact Process, sulphur trioxide is formed. Sulphur trioxide would react with water to form sulphuric acid, as shown in this reaction: $SO_{3\,(g)} + H_2O_{(l)} \rightarrow H_2SO_{4\,(aq)}$

Instead, sulphur trioxide is dissolved in sulphuric acid to form fuming sulphuric acid, $H_2S_2O_7$. Why is this done?

Reacting sulphur trioxide with water is very exothermic. The H_2SO_4 would evaporate making the process unworkable.

(2 marks)

(d) How is the sulphur dioxide used for the Contact Process formed?

Sulphur or metal sulphides are roasted in air.

(1 mark)

2   (a) How are free radicals such as Cl· formed?

A covalent bond breaks so that each side keeps one of the shared electrons. Usually caused by UV light.

(2 marks)

(b) Why is the chlorine free radical Cl· more reactive than the chloride ion Cl⁻?

It has an unpaired electron. Cl⁻ doesn't. Unpaired electrons are very unstable.   *Paired electrons are much more stable than unpaired electrons.*

(3 marks)

Exam Questions

1   The Contact Process is used in the synthesis of sulphuric acid. The process involves this reaction.

$$2SO_{2(g)} + O_{2(g)} \rightleftharpoons 2SO_{3(g)} \qquad \Delta H = \text{-297 kJ Mol}^{-1}.$$

The conditions are as follows: 450°C, 1-2 atm pressure and a vanadium pentoxide catalyst.
These conditions are a compromise between yield and operating costs.

(a) (i) What pressure conditions would give a greater yield of sulphur trioxide?

....

(1 mark)

(ii) Why are these pressure conditions not used?

....

....

(2 mark)

(iii) What temperature conditions would give a greater yield of sulphur trioxide?

....

(1 mark)

(iv) Why are these temperature conditions not used?

....

....

(2 marks)

(b) How does the vanadium pentoxide catalyst help the reaction?

....

(1 mark)

2   The protective ozone layer in the Earth's atmosphere can be damaged by free radicals.
A chlorine free radical can react with an ozone molecule (O_3) as shown by this equation:

$$Cl\cdot + O_3 \rightarrow ClO + O_2$$

(a) Explain how one free radical can destroy up to one million molecules of ozone.

....

....

(1 mark)

(b) Chlorofluorocarbons (CFCs) can break down into free radicals in the upper atmosphere.
Why does this happen in the upper atmosphere, and not at ground level?

....

....

(2 marks)

(c) Why have some countries of the world banned the use of CFCs?

....

(1 mark)

Homologous Series and Isomers

The rest of this section is about <u>organic chemistry</u>, which is largely to do with 'families' of <u>carbon compounds</u>. The members of each of these families are all pretty <u>similar</u>.

Homologous Series are Families of Carbon Compounds

The members of a homologous series all have certain things <u>in common</u>.

1) They all have the <u>same</u> general <u>formula</u>.

2) They all have <u>similar reactions</u>.

3) They can all be <u>made</u> in <u>similar</u> ways.

> <u>Alkanes</u> and <u>alkenes</u> are examples of <u>homologous series</u>. Alkanes have a general formula of C_nH_{2n+2}, and alkenes have a general formula of C_nH_{2n}.

4) Certain <u>properties</u> change <u>gradually</u> as you go through the series (i.e. as more carbon atoms are added). For example, as the molecules get <u>larger</u>, the <u>melting</u> and <u>boiling</u> points <u>increase</u>.

> **The <u>names</u> of the members of many homologous series follow a similar <u>pattern</u>.**
> <u>Alkanes</u>: <u>methane</u> (CH_4), <u>ethane</u> (C_2H_6), <u>propane</u> (C_3H_8) and <u>butane</u> (C_4H_{10}).
> <u>Alkenes</u>: <u>ethene</u> (C_2H_4), <u>propene</u> (C_3H_6) and <u>butene</u> (C_4H_8).

Things starting with 'eth-' have 2 carbon atoms, 'prop-' have 3, and so on.

The Functional Group gives a Family its Character

1) The <u>functional group</u> is the <u>characteristic</u> that all members of a homologous series have <u>in common</u>. It's the functional group (or lack of one, in the case of alkanes) that's responsible for each member of that family <u>acting</u> in a <u>similar way</u>.

2) So for <u>alkenes</u>, the functional group is the pair of <u>double-bonded carbon atoms</u>.

3) <u>Alcohols</u> (see page 223) have the functional group <u>-OH</u>.

4) And <u>carboxylic acids</u> (see page 225) have the functional group <u>-COOH</u>.

Isomers have their Atoms Arranged Differently

Organic chemistry is a <u>huge</u> subject. One reason is that the atoms in many compounds can be <u>arranged</u> in <u>different ways</u>. Compounds which have the <u>same formula</u> but <u>different structures</u> are called <u>isomers</u>.

Example

<u>Pentane</u> has the formula C_5H_{12}, but the atoms in pentane can be arranged as below:

Isomer 1

H H H H H
| | | | |
H-C-C-C-C-C-H
| | | | |
H H H H H

Isomer 2

H H H H
| | | |
H-C-C-C-C-H
| | | |
H H C H
|
H - H
|
H

Isomer 3

H
|
H H-C-H H
| | |
H-\C-C-C/-H
H H-C-H H
|
H

1) Different <u>isomers</u> often have different <u>physical properties</u> (so things like <u>melting</u> and <u>boiling points</u> might be different, for example).

2) These different properties are because the <u>strength</u> of the <u>intermolecular bonds</u> depends on the exact <u>arrangement</u> of the atoms.

3) Molecules that are <u>long</u> and <u>thin</u> have <u>higher</u> melting and boiling points than <u>branched</u> molecules.

4) This is because long, thin molecules <u>pack together</u> better, and so have more <u>contact</u> with other molecules. This means that the intermolecular forces are <u>stronger</u>.

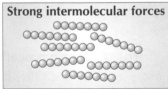
Strong intermolecular forces

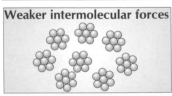
Weaker intermolecular forces

Don't confuse isomers with isotopes

One good thing about organic chemistry is that it's all about learning patterns and trends. Learn the basics (like the <u>meth-</u>, <u>eth-</u>, <u>prop-</u>, <u>but-</u> names), and you can keep using the same facts, time and time again.

Alcohols

Alcohols are carbon chains with an -OH functional group. It's the -OH which makes them alcohols.

Alcohols have an '-OH' Functional Group and end in '-ol'

The basic naming system is the same as for alkanes — but replace the final '-e' with '-ol'.

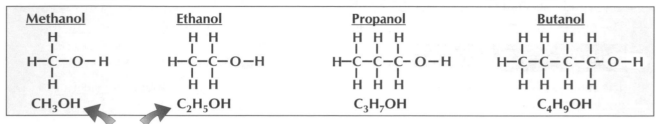

Methanol	Ethanol	Propanol	Butanol

CH_3OH C_2H_5OH C_3H_7OH C_4H_9OH

Don't write CH_4O instead of CH_3OH, or C_2H_6O instead of C_2H_5OH, etc., as it doesn't show the functional -OH group.

All Alcohols react in a Similar Way

Being members of a homologous series, the reactions of all alcohols are similar.

1) Combustion

All alcohols burn, giving carbon dioxide and water (and heat).

$$2CH_3OH + 3O_2 \longrightarrow 2CO_2 + 4H_2O \qquad C_2H_5OH + 3O_2 \longrightarrow 2CO_2 + 3H_2O$$

2) Reaction with sodium

Alcohols react with sodium metal to give off hydrogen.

$$2C_2H_5OH + 2Na \longrightarrow 2C_2H_5ONa + H_2$$

3) Esterification

Alcohols react with carboxylic acids (see page 225) to form esters (see page 226).

Alcohols are Used as Fuels and Solvents

1) Some alcohols are used as fuels. Ethanol is used as a fuel in 'spirit' burners — it burns cleanly, is non-smelly and washes off with water. In fact, some countries that have little or no oil deposits but plenty of sunshine (e.g. Brazil) grow loads of sugar cane, which they ferment to form ethanol (see page 224). This can then be added to petrol.

2) Ethanol (as well as some other alcohols) is used as a solvent. 'Methylated spirit' (or 'meths') is mainly ethanol, but it has methanol added to it to make it 'undrinkable', and a blue dye to stop people drinking it by mistake. It's used to clean paint brushes.

Cholesterol contains the Alcohol Group -OH

1) Cholesterol contains the alcohol functional group -OH.
2) It belongs to the group of compounds known as steroids.
3) Cholesterol is essential to your body chemistry. Your body produces it, but if excess amounts are present, it's thought to contribute to heart disease.

OH — It's all about alcohols...

This is a nice page to learn. There are a few equations, but nothing you can't handle. As for the facts about the uses of alcohol, you probably know most of those already. Still, make sure you learn all this.

Ethanol

Make Ethanol by **Fermentation** or **Hydration of Ethene**

1) Fermentation

Ethanol has been made by fermentation for thousands of years.

1) The process of fermentation converts sugars (usually from fruits and vegetables) into ethanol.

$$C_6H_{12}O_6 \longrightarrow 2C_2H_5OH + 2CO_2$$

 See pages 161-164 of this book, and the Biology book for more info about enzymes.

2) The reaction is brought about by enzymes (naturally occurring catalysts) found in yeasts, and goes quickest at about 30 °C (if it gets too hot the enzymes are destroyed).

3) During fermentation, it's important to prevent oxygen getting at the alcohol. This is because oxygen converts ethanol to ethanoic acid (which is what you get in vinegar). It's also because if oxygen is present, the yeast respires aerobically to produce just carbon dioxide and water.

4) When the concentration of alcohol reaches about 10-15%, the reaction stops, due to the yeast being killed by the alcohol. (Spirits are made by distilling already fermented brews.)

2) Hydration of Alkenes

This is how alcohols are commonly made industrially.

Ethene, made during the cracking of crude oil (see Section Two), will react with steam at 300 °C and a pressure of 70 atmospheres, using phosphoric acid as a catalyst. The product is ethanol.

These Methods have their **Pros** and **Cons**

Fermentation	Hydration of Ethene
1. **Low tech**	1. **Requires large chemical plant**
2. Uses a **renewable** resource	2. Ethene is a **finite** resource
1. **Expensive** to concentrate and purify 2. Fermentation is **slow** 3. Made in **batches** (i.e. process can't be left to run continuously, so less efficient)	1. **High quality** product 2. **Quick** 3. **Efficient**, as process runs **continuously**

Ethanol is used to make Ethene and in Alcoholic Drinks

To make ethene
i) The plastics and polymer industries use lots of ethene.

ii) Countries which have no oil but plenty of land for growing crops for fermentation can make ethene through the dehydration of ethanol, using a heated aluminium oxide catalyst. (The reaction is the exact opposite of the one above illustrating the Hydration of Alkenes.)

In drink Alcoholic drinks contain ethanol. Beers contain around 3-6% ethanol, wines about 9-15%. Spirits like whisky, gin and brandy are made by the fractional distillation of wines and other fermented brews. The distilled liquid has a higher percentage of alcohol, usually around 40%.

Alkenes? Surely alcohol comes from grapes

At the top it says you make ethanol from ethene, and then later it says you make ethene from ethanol. It all depends on what you've got more of — if you've got lots of one thing, you can make it into the other.

Carboxylic Acids

Carboxylic Acids have Functional Group -COOH

1) Carboxylic acids have '-COOH' as a functional group.

2) They are <u>weak</u> acids — less than 1% of the molecules <u>ionise</u> in water.

> e.g. ethanoic acid: $CH_3COOH(l) \rightleftharpoons CH_3COO^-(aq) + H^+(aq)$

3) Their names end in '-<u>anoic acid</u>' (and start with the normal '<u>meth/eth/prop/but</u>').

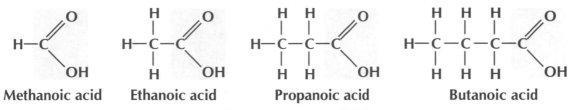

Methanoic acid   **Ethanoic acid**   **Propanoic acid**   **Butanoic acid**

Carboxylic Acids react like other Acids

Carboxylic acids... well, they're <u>acids</u>. And they do normal <u>acid things</u>:

1) With <u>metals</u> they form <u>salts</u> and <u>hydrogen</u>:

The salts are <u>ethanoates</u> — e.g. magnesium ethanoate.

> $2CH_3COOH + Mg \longrightarrow H_2 + Mg^{2+}(CH_3COO^-)_2$

2) With <u>carbonates</u> (CO_3^{2-}) and <u>hydrogencarbonates</u> (HCO_3^-) they form <u>salts</u>, <u>carbon dioxide</u> and <u>water</u>:

> $Na_2CO_3 + 2CH_3COOH \longrightarrow 2Na^+CH_3COO^- + CO_2 + H_2O$
>
> $Mg(HCO_3)_2 + 2HCOOH \longrightarrow Mg^{2+}(HCOO^-)_2 + 2CO_2 + 2H_2O$

3) With <u>alkalis</u> they are <u>neutralised</u>:

> $H^+ + OH^- \longrightarrow H_2O$

This is the <u>ionic equation</u> for a <u>neutralisation</u> reaction — the <u>H⁺</u> represents the acid, the <u>OH⁻</u> represents the <u>alkali</u>.

4) With <u>indicators</u> they give typical <u>acid</u> colours:

> Universal indicator goes orange / red.   Phenolphthalein is colourless.

Some Carboxylic Acids are Fairly Common

1) <u>Ethanoic acid</u> is the acid in <u>vinegar</u>, which is used for <u>flavouring</u> and <u>preserving</u> foods.

2) If <u>wine</u> or <u>beer</u> is left open to the <u>air</u>, the <u>ethanol</u> is <u>oxidised</u> to <u>ethanoic acid</u>. This is why drinking wine after it's been open for a couple of days is like drinking vinegar — basically it <u>is</u> vinegar.

> ethanol + oxygen $\longrightarrow$ vinegar + water
>
> $CH_3CH_2OH + O_2 \longrightarrow CH_3COOH + H_2O$

3) Ethanoic acid is also used in the manufacture of <u>rayon</u>, a synthetic fibre.

4) <u>Citric acid</u> is present in <u>oranges</u> and <u>lemons</u>, and is manufactured in large quantities to make fizzy drinks. It's also used as a <u>descaler</u> (see page 203 about hard water).

5) <u>Aspirin</u> is a <u>man-made</u> carboxylic acid. It's not only an <u>analgesic</u> (painkiller), but it reduces <u>blood clotting</u> slightly, and is regularly taken by people at risk of <u>heart attack</u>.

COOH — what a fantastic page about acids...

You should also know a little about <u>ascorbic acid</u> (<u>vitamin C</u>). Ascorbic acid is in loads of green vegetables, oranges and tomatoes, and so on. It's very important for <u>human health</u> in loads of ways.

You don't need to learn this page if you're doing the OCR syllabus.

Esters

It's the topic you've been waiting for... <u>esters</u>. This page may not change your life exactly, but by the time you've read it through to the end, you'll know what makes those nice <u>pear-drop</u> sweets smell.

Alcohol + Acid ⟶ Ester + Water

1) <u>Esters</u> are formed from an <u>alcohol</u> and a <u>carboxylic acid</u>. This process is called <u>esterification</u>.

2) An <u>acid catalyst</u> is usually used (e.g. concentrated <u>sulphuric acid</u>).

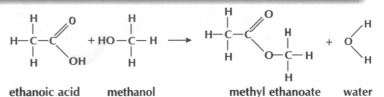

CH_3COOH + C_2H_5OH ⟶ $CH_3COOC_2H_5$ + H_2O

Ethanoic acid   +   **Ethanol**   ⟶   **Ethyl ethanoate**   +   **Water**

Learn what you get with other Alcohols and Carboxylic Acids

1) The <u>alcohol</u> forms the <u>first</u> part of the ester's name.

2) The <u>acid</u> forms the <u>second</u> part.

ethanoic acid   **methanol**   **methyl ethanoate**   **water**

Esterification is Reversible

1) The above process is <u>reversible</u> — an ester reacts with water to form an <u>alcohol</u> and a <u>carboxylic acid</u>.

2) This process is called <u>ester hydrolysis</u>.

Esters are often used in Flavourings and Perfumes

1) Esters have quite <u>strong smells</u>, though these are usually quite <u>pleasant</u>.

2) Esters are pretty common in <u>nature</u>. Loads of common <u>food smells</u> (plus those in products like <u>perfumes</u>) contain natural esters.

3) Esters are also manufactured <u>synthetically</u> to <u>enhance</u> food <u>flavours</u> or <u>aromas</u>, e.g. there are esters that smell of rum, apple, orange, pineapple, and so on. And esters are responsible for the distinctive smell of <u>pear-drops</u>.

4) Other esters are the basis of familiar smells in drinks, glues, paints and ointments (e.g. that lovely Deep-Heat smell).

5) And the formation of natural esters in <u>wines</u> and <u>spirits</u> gives a pleasant aroma to these drinks, and <u>removes</u> a lot of the <u>acids</u> (that are partly responsible for the <u>hangover</u> headaches).

I bet you never knew Chemistry could smell so good

Now you know the answer to the burning question of what makes <u>pear-drops</u> smell the way they do. It's all down to <u>esters</u>. And <u>polyesters</u> (of course) are things made from lots of ester molecules.

Warm-Up and Worked Exam Questions

With organic chemistry, it's important to get the names of the molecules right. Make sure you know your propanol from your propanoic acid, and your methyl butanoate from your butyl methanoate.

Warm-up Questions

1) What is the name for two molecules with the same formula but different molecular structure?
2) How does the boiling point vary with the amount of chain branching in an alkane?
3) Methylated spirit is mostly ethanol with some methanol. Why is methanol added?
4) Are carboxylic acids weak acids or strong acids?
5) What is the product of the reaction between butanol and methanoic acid?

Worked Exam Questions

These worked exam questions give you a good idea of what you could be asked in the Exam.

1   This diagram shows propanoic acid.
        One part of the molecule is responsible for its reactions as an acid.

It's the COOH bit.

    a)   Circle the part of the molecule responsible for propanoic acid's reactions.

(1 mark)

    b)   Would you expect to find that part of the molecule in ethanoic acid and butanoic acid?

     Yes   *Because the carboxylic acids are a homologous series.*

(1 mark)

    c)   What is the name for the part of a molecule that is responsible for the similarities in chemical properties between members of a homologous series (e.g. the carboxylic acids)?

     A Functional Group

(1 mark)

2   Look at these two isomers of hexane.

A

B

    a)   Which molecule do you think will have the higher boiling point? Explain your answer.

     A, because it is less branched than B. Molecules with less branching
     have stronger intermolecular forces, which require more energy to
     break. The less branching, the stronger the bonds.

(2 marks)

    b)   Which molecule is 2,3-methylbutane?   *4 carbon atoms in the main chain because it's butane,*
                                                and two methyl groups on the 2nd and 3rd carbons.

     B

(1 mark)

Exam Questions

1   The alcohols form a homologous series. They all have similar chemical properties.

a)   Write a balanced equation for the combustion of methanol.

....
(1 mark)

b)   Explain why ethanol makes a clean, cheap fuel.

....

....

....
(2 marks)

c)   Alcohols react with sodium.

   i)   Write out a balanced equation for the reaction between ethanol and sodium.

....
(3 marks)

   ii)   What other reaction of sodium is this similar to?

....
(1 mark)

2   This question is about the manufacture of ethanol.

a)   Ethanol is traditionally made by the fermentation of sugar.

   i)   Complete and balance this equation for the fermentation of glucose.

   $C_6H_{12}O_6 \rightarrow$ .... + ... CO_2
(3 marks)

   ii)   The carbon dioxide forms a raft of foamy bubbles on the surface, which prevents oxygen getting at the mixture. Why is it important to prevent oxygen from getting at the fermenting mixture?

....

....
(2 marks)

b)   Ethanol can be produced industrially by a different method to fermentation.

   i)   What are the raw materials used in this process?

....
(2 marks)

   ii)   Under what conditions is this process carried out?

....
(2 marks)

Exam Questions

3   Carboxylic acids and esters are found naturally in some foods.

a)   In what common foodstuff is ethanoic acid found?

....
(1 mark)

b)   Wine goes sour when it is left out in an open bottle.
Write a word equation to show what happens.

....
(1 mark)

c)   What is the common name for the carboxylic acid found in lemons?

....
(1 mark)

d)   Esters are responsible for the distinctive smells of some foods. Give an example of one of these foods.

....
(1 mark)

4   a)   What are the products of reactions between alcohols and carboxylic acids?

....
(1 mark)

b)   Fiona mixes methanoic acid with ethanol in a test tube. She adds a couple of drops of a catalyst, and then pours the mixture into a beaker of water.

i)   What is the name of the main product of this reaction?

....
(2 marks)

ii)   What could Fiona use as a catalyst?

....
(1 mark)

iii)   Write down what you think Fiona will notice when she smells the mixture in the beaker?

....
(1 mark)

c)   Complete this diagram which shows the reaction between methanoic acid and ethanol.

$$H-C{\overset{\displaystyle O}{\underset{\displaystyle OH}{<}}} \; + \; H\text{-}O\text{-}\underset{\underset{\displaystyle H}{|}}{\overset{\overset{\displaystyle H}{|}}{C}}\text{-}\underset{\underset{\displaystyle H}{|}}{\overset{\overset{\displaystyle H}{|}}{C}}\text{-}H \; \rightarrow$$

(1 mark)

Carbohydrates

The three major nutrients in food are <u>carbohydrates</u>, <u>proteins</u> and <u>fats</u>.
And it's the <u>carbohydrates</u> that give us most of our <u>energy</u>.

*Carbohydrates — made of **Carbon, Hydrogen** and **Oxygen***

1) Carbohydrates consist of <u>carbon</u> (C), <u>hydrogen</u> (H) and <u>oxygen</u> (O).
   They all have the general formula:

$$C_x(H_2O)_y$$

 There are always twice as many hydrogens as there are oxygens.

2) <u>Glucose</u>, <u>sucrose</u>, <u>starch</u> and <u>cellulose</u> are all carbohydrates.

Two **Monosaccharides** *Join to make a* **Disaccharide**

1) Small soluble carbohydrates are called <u>sugars</u>, e.g. <u>glucose</u>.

2) Glucose is a <u>monosaccharide</u> — it's made of a
   <u>single</u> sugar unit — this sugar unit is a <u>ring-shaped</u> molecule.

3) When <u>two</u> monosaccharides bond together a
   <u>disaccharide</u> is formed — it has two rings.

4) The monosaccharides glucose and fructose join to form <u>sucrose</u> —
   a disaccharide. This is called a <u>condensation</u> reaction because a
   <u>water</u> molecule is produced as the two molecules join.

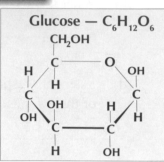

Glucose — $C_6H_{12}O_6$

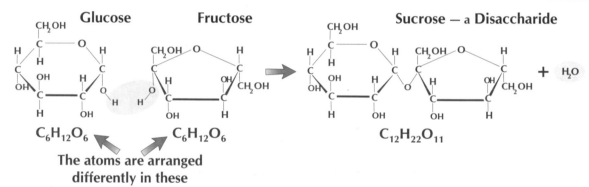

Glucose   Fructose   Sucrose – a Disaccharide

$C_6H_{12}O_6$ $C_6H_{12}O_6$ $C_{12}H_{22}O_{11}$ + H_2O

**The atoms are arranged
differently in these**

Polysaccharides are *Polymers* of Monosaccharides

1) <u>Monosaccharides</u> join together in long <u>chains</u> to form <u>polysaccharides</u>.

2) Polysaccharides form by <u>condensation polymerisation</u> — a <u>water</u> molecule is
   produced for every monosaccharide that joins on.

3) <u>Starch</u> and <u>cellulose</u> are polysaccharides — they are polymers of <u>glucose</u>.

4) You need to be able to <u>draw</u> the outline structural formulae for starch and cellulose.
   Starch is <u>easy</u> to draw — just a simple chain of glucose molecules.
   Cellulose is pretty similar, except <u>alternate</u> glucose molecules are <u>flipped</u> over.

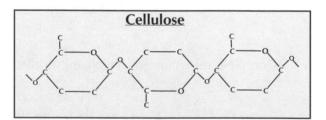

Starch   **Cellulose**

Carbohydrates are exam staples

The worst thing about this page is all the <u>structures</u> you need to learn — cover the page and draw
<u>glucose</u>, <u>sucrose</u>, <u>starch</u> and <u>cellulose</u>. Keep trying 'til you get them spot on.

Proteins and Fats

Amino acids are pretty useful — they make up proteins, which we need to make new cells and to grow.

Amino Acids *have a* -COOH *Group and an* -NH₂ *Group*

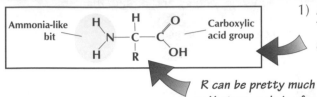

Ammonia-like bit

Carboxylic acid group

1) Amino acids are easy to recognise. They always contain a carboxylic acid group -COOH and an ammonia-like bit -NH₂ (an amine group).

R can be pretty much anything — an H atom, a chain of carbon atoms...

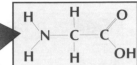

2) The simplest amino acid is glycine — it has two hydrogens on the central carbon.

Proteins are *Polymers* of *Amino Acids*

1) Ammonia reacts with acids — so the ammonia-like bit at the end of one amino acid molecule will react with the carboxylic acid bit at the end of another amino acid molecule.

Peptide links join one amino acid to the next — that's why the chains are called polypeptides.

2) In this way, loads of amino acids join together to form a polymer. Polymers of amino acids are called polypeptides or proteins.

3) This reaction is called condensation polymerisation because a water molecule forms as the amino acids join together. This is just like the formation of polysaccharides (see page 230).

Peptide link

Polymers are long chains of small molecules.

Triglycerides are Esters made from *Glycerol*

1) Triglycerides are fats or oils.

2) They're esters made from glycerol — an alcohol with 3 -OH groups.

Example

Glycerol

Carboxylic acid

R is a long carbon chain here

Triglyceride

See pages 223-226 for more about alcohols, carboxylic acids and esters.

Big Molecules can be *Broken Down* by *Hydrolysis*

1) Reactions which form polypeptides, triglycerides and polysaccharides (see page 230) also form water.

2) These reactions are all reversible. With water, the big molecules slowly revert to small molecules.

Hydrolysis (with water)

Polypeptides ⟶ Amino acid

Triglycerides ⟶ Glycerol + carboxylic acid

Polysaccharides ⟶ Monosaccharides

Proteins and fats are both made up of smaller molecules

Make sure you can spot an amino acid from a mile away, even if it's cunningly disguised as an alcohol molecule. As for the triglyceride molecule — you'd better get on with memorising that one.

Diet

A *Balanced* Diet is Important for *Good Health*

1) A balanced diet provides all the stuff our bodies need. Different people need different diets.

2) The more active you are the more energy you need. Children need more energy and protein than an adult of the same size would need. Adolescents are growing quickly and need lots of food too. Pregnant and breast-feeding women need extra food for the baby as well as themselves.

3) A balanced diet must contain the correct amounts of the following:

Fats and *Carbohydrates* provide Energy

1) Fats and carbohydrates provide energy to move and keep warm.

2) Too much sugar (a carbohydrate) leads to tooth decay.

3) Fats can be saturated or unsaturated.

4) Unsaturated fats come from plants and have C=C double bonds in their carbon chains.

5) Saturated fats come mainly from animals and have no C=C double bonds. Eating excessive saturated fat increases the risk of heart disease.

Making Margarine

1) Unsaturated vegetable oils are used to make margarine.

2) A nickel catalyst can make some double bonds break and accept hydrogen atoms. The oil then becomes a solid fat (hydrogenated vegetable oil).

$$-(C=C)- + H_2 \rightarrow -(\overset{H}{C}-\overset{H}{C})-$$

Protein — for Energy and Growth

1) Protein's needed for growth, and cell repair and replacement.

2) You need to eat certain amino acids so that your body can make other amino acids it needs.

3) Vegetarians must be careful to eat enough protein to make sure they get the amino acids which our bodies can't make.

Fibre — Indigestible Carbohydrate

1) Fibre is mainly cellulose from plants.

2) It keeps food moving smoothly through your digestive system by providing something for your gut muscles to push against.

Vitamins and *Minerals* are needed in Tiny Amounts

Vitamin A: For making pigment in the eye.
Vitamin C: Keeps the skin strong and supple. Prevents scurvy (a nutritional disease). Protects tissues by destroying harmful free radicals (see page 219). Vitamin C Levels are reduced by storing food for too long or overcooking it.
Vitamin D: Needed for strong bones and teeth.

Calcium: For healthy bones and teeth.
Iron: For making haemoglobin in red blood cells.

Raising Agents make Bread and Cakes Rise when Cooked

1) Cakes and bread rise when cooked — gases trapped inside the ingredients expand when heated.

2) When baking bread, carbon dioxide is produced by yeast fermenting the sugar (see page 224).

3) Baking soda is a mixture of sodium hydrogencarbonate and tartaric acid. It releases carbon dioxide when it becomes wet.

Additives are *Chemicals* added to Food

1) The E-number system is a list of permitted food additives. Some are natural but most are man-made.

2) Some additives don't have any effect on you, e.g. E175 — gold, while some are beneficial, e.g. E300 — vitamin C. Others can be harmful, e.g. salt contributes to high blood pressure. Some people try to avoid foods with certain additives in.

Colourings — Improve appearance of food, e.g. E160b (Annatto, a yellow dye).
Flavourings — Improve taste, e.g. MSG.
Preservatives — Make food last longer, e.g. E220, Sulphur dioxide.
Sweeteners — Food can taste sweet without being so fattening, e.g. Nutrasweet.

Foods are chemicals too

You need to know what you need and why you need it. Make sure you know the difference between unsaturated fat and the artery-clogging, saturated kind. You know the drill — learn the headings, then test yourself.

Drugs

Drugs aren't just things you're told to say no to — they're anything which <u>alters</u> your <u>body's chemistry</u>. Drugs can be <u>dangerous</u>, but some cure people of conditions which would have killed them in the past.

Drugs affect **Chemical Reactions** in the Body

1) Drugs are things we <u>take</u> which change <u>chemical reactions</u> in the body.

2) For example, <u>analgesics</u> are drugs used to reduce <u>pain</u>, e.g. aspirin, paracetamol and ibuprofen.

Aspirin isn't a **Recent** Invention

1) Aspirin is used to relieve <u>pain</u> and reduce <u>fever</u>. It can also stop <u>blood clots</u> forming, and so reduce the danger of <u>heart attacks</u> and <u>strokes</u>.

2) The Greeks used <u>white willow bark</u> to reduce pain — the problem was it also caused vomiting. The painkilling ingredient was <u>salicylic acid</u> — and in 1893, a German scientist made a similar chemical which had fewer <u>side effects</u>. This drug is known as <u>aspirin</u>.

3) Aspirin is now the <u>largest-selling</u> drug in the world.

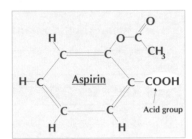

Soluble Aspirin Works **Faster** than Normal Aspirin

1) Aspirin works at the <u>site</u> of an injury by stopping <u>prostaglandin</u> being made. Prostaglandin is a chemical which causes <u>swelling</u> and is involved in the feeling of <u>pain</u>.

2) Aspirin molecules are <u>not</u> very soluble so they get to the injury <u>slowly</u>.

3) But since aspirin is an <u>acid</u>, it reacts with <u>sodium carbonate</u> to form a <u>soluble</u> salt. This salt is the <u>soluble</u> form of aspirin and gets into the blood more <u>quickly</u>.

New Drugs are Extensively **Tested**

1) Modern drugs now <u>cure</u> illnesses that were once fatal — but there are <u>still</u> diseases with <u>no</u> known cure.

2) Constant research is done to find <u>new</u> and <u>better</u> drugs.

3) <u>Extensive testing</u> is done on potential new drugs in <u>controlled experiments</u> with many <u>stages</u>. <u>Only</u> drugs which pass one stage continue to the next.

STAGES OF DRUG TESTING

1) The drug is tested on <u>cells</u> growing in a tissue culture.
2) Next it is tested on laboratory <u>animals</u>, such as rats — many people object to animal testing.
3) If the drugs have the desired effect, <u>human</u> trials start.
    Phase 1 — A <u>small</u> group (less than 100) volunteers.
    Phase 2 — A <u>larger</u> group of patients (several hundred).
    Phase 3 — A <u>very large</u> group of patients (tens of thousands).

A patient can sometimes get better because they believe they're being treated (even when they're not). A placebo is used as a comparison to the drug being tested.

4) At each stage extensive <u>observation</u> and <u>recording</u> of results is done.

5) The results are <u>compared</u> against established drugs and <u>placebos</u> (fake drugs that have no <u>physical effect</u>) in <u>double blind tests</u> — neither the <u>patients</u> or <u>researchers</u> know who took which pills until later.

6) A recent trial was <u>stopped</u> when it became obvious that the drug being tested was having a positive effect — it was thought <u>wrong</u> to carry on using placebos when a successful drug had been found.

THE LAST PAGE
That's it. If you've learnt it all properly you should know chemistry by now. Just one more set of questions to whistle through, and then you're ready to try an Exam. Go get 'em.

Warm-Up and Worked Exam Questions

Most of the stuff about food and aspirin is relatively basic. The difficult part is where you might be asked to sketch a structural diagram of a food molecule, or explain how two molecules polymerise.

Warm-up Questions

1) What three kinds of compound are needed by the body for energy?
2) What is the formula of glucose?
3) What molecules join together to make protein?
4) Name a raising agent used to make bread or cakes rise.
5) What type of nutrient are triglycerides?
6) What drug contains salicylic acid?

Worked Exam Question

This is a fairly straightforward question about carbohydrates. Take notice of the part about condensation polymerisation — it can come up in questions about protein, too.

1   This question is about the polymerisation of simple sugars.

A molecule of sucrose is formed when a molecule of glucose joins onto a molecule of fructose.

a)   Write a word equation to show the reaction between glucose and fructose.

glucose + fructose → sucrose + water

Don't forget the water! ⟶

(1 mark)

b)   Glucose is a monosaccharide. What is sucrose?
   *Two sugar units =
   Disaccharide*

   Sucrose is a disaccharide ⟵

(1 mark)

c)   Many monosaccharide molecules can polymerise to make a long chain polysaccharide.

   i)   This polymerisation is called condensation polymerisation. Why is this?

   A molecule of water is formed for each monosaccharide

   that joins the chain.   *Imagine the water condensing.*
                                  That's all it is.

(1 mark)

   ii)   Starch is a polysaccharide.
        What kind of organisms make starch, and what do they use it for?

   Plants make starch and use it to
                          One mark for "plants",
   store food.   *one mark for "store food".*

(2 marks)

   iii)   Amylase is an enzyme which breaks down starch into its individual units.
         What substance is produced when starch is broken down by amylase?

   Glucose   *You'll know this from Biology lessons.*

(1 mark)

Exam Questions

1   This diagram shows glycine. Glycine is an amino acid. It contains an amine functional group, and a carboxylic acid functional group.

$$H_2N-\overset{\overset{\displaystyle H}{|}}{\underset{\underset{\displaystyle H}{|}}{C}}-C\overset{\displaystyle \diagup O}{\diagdown OH}$$

glycine

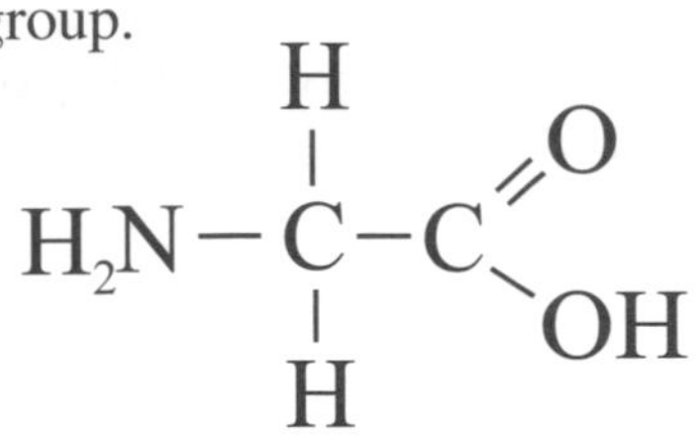

a) One molecule of glycine can react with another.

   i)   Which part of the glycine molecule reacts with the amine group on the other molecule?

   ....
   (1 mark)

   ii) William says that joining amino acids together is an example of condensation polymerisation. Faith says that condensation polymerisation only refers to joining monosaccharides together. Who is right? Explain your answer.

   ....

   ....

   ....
   (3 marks)

b) Amino acids polymerise to form long chains.

   i)   What substances are produced when amino acids polymerise.

   ....
   (1 mark)

   ii) The polymerisation reaction is reversible. What is the name of the reverse reaction?

   ....
   (1 mark)

2   Vitamins and minerals are part of a balanced diet.

a) Give an example of a mineral required by the body, and say what it is needed for.

   ....
   (1 mark)

b) In the 18th century, sailors on long voyages suffered from a disease called scurvy.
    They were given limes in their rations to prevent scurvy. Scurvy is caused by a lack of which vitamin?

   ....
   (1 mark)

Exam Questions

3   a)   Most fats and oils are triglycerides.

       i)   Draw a diagram to show the structure of a glycerol molecule.

(2 marks)

       ii)   What substances react with glycerol to form triglycerides?

....

(1 mark)

   b)   Butter contains a very high proportion of saturated fat. Margarine made from sunflower oil contains a high proportion of unsaturated fat.

       i)   Explain what is meant by the terms saturated fat and unsaturated fat.

....

....

(2 marks)

       ii)   Which is considered to be more healthy for the body, saturated fat or unsaturated fat?

....

(1 mark)

4   Aspirin is a drug used to treat pain.

   a)   Heart patients are sometimes advised by their doctors to take one junior aspirin every day. Explain why doctors give this advice.

....

....

(1 mark)

   b)   Potential new drugs are tested in controlled conditions.

       i)   Why are new drugs tested on tissue cultures and on rats before human testing?

....

(1 mark)

       ii)   New drugs are tested against a placebo. What is a placebo and why are they used?

....

....

(2 marks)

Revision Summary

These questions are all perfectly tailored to help you check what you know and what you don't from the last section.

1) Why might it be beneficial to get a thicker oxide layer on the surface of aluminium?

2) Name the electrolyte used in the process of "anodising" aluminium.

3) Is titanium above or below carbon in the reactivity series?
Why isn't titanium oxide reduced with carbon?

4) Balance the equation for the reduction of titanium chloride with magnesium:
$Mg + TiCl_4 \rightarrow MgCl_2 + Ti$.

5) Draw a table to compare the composition, main properties and main uses of the following metals:
cast iron, high carbon steel, mild steel and stainless steel(s).

6) In the first stage of producing steel, what is removed from the molten iron by blowing oxygen through it? Explain how this works.

7) In order to plate a copper item with nickel in an electrolysis cell, what should be used for
a) the anode, and b) the cathode? c) Name a suitable electrolyte.

8) What is the "Contact Process" used to make? Describe the process in detail, including equations.

9) What is a dehydrating agent? What are the products of the reaction between concentrated sulphuric acid and sugar?

10) How many free (unpaired) electrons does a free radical have?

11) What were (or are) chlorofluorocarbons used for?

12) What are 'isomers'? Draw three isomers of pentane, C_5H_{12}.

13) Why is too much cholesterol bad for you?

14) Describe two commercial ways of making ethanol.
List two advantages and two disadvantages of each method.

15) What is the functional group of carboxylic acids?
Draw the structure of the first four carboxylic acids.

16) Write an equation for ethanoic acid reacting with a) magnesium, and b) sodium hydrogencarbonate.

17) Draw the structure of methyl ethanoate. Write an equation for its formation.

18) State one important property of esters.

19) Give four examples of carbohydrates.

20) Name and draw a monosaccharide. Name and draw a disaccharide.

21) What two chemical groups do all amino acid molecules have?

22) What type of reaction occurs when amino acids join together? Why is it called this?

23) What is hydrolysis? Name three types of compound that undergo hydrolysis.

24) How are vegetable oils converted into margarine?

25) Why do vegetarians have to be more careful about their diet than non-vegetarians?

26) Why do we need the following?
a) Vitamin A   b) Vitamin C   c) Vitamin D   d) Calcium   e) Iron

27) What's another word for 'painkiller'? Name three common painkillers.

28) Draw a molecule of aspirin.

29) Why does plain aspirin take so long to start working? What is sometimes used instead?

30) Describe the three main phases of drug testing.

Practice Exam

Once you've been through all the questions in this book, you should feel pretty confident about the exam. As final preparation, here is a **practice exam** to really get you set for the real thing. It's split into **two sections** — one for if you're doing Double Science and an extra section for if you're taking GCSE Chemistry (i.e. Separate Sciences). The paper is designed to give you the best possible preparation for the differing question styles of the actual exams, whichever syllabus you're following. If you're doing Foundation then you won't have learnt every bit — but it's still good practice.

CGP — Practice Exam Paper / GCSE Double Science / GCSE Chemistry

General Certificate of Secondary Education

GCSE Science: Double Award
GCSE Chemistry

Centre name					
Centre number					
Candidate number					

Paper 1

Surname	
Other names	
Candidate signature	

Time allowed:

GCSE Science: Double Award

Section A	1 hour 15 minutes

GCSE Chemistry

Sections A and B	1 hour 45 minutes

Instructions to candidates

- If you are taking **GCSE Science: Double Award** answer **all** of the questions in **Section A**.
  Do **not** answer Section B.
- If you are taking **GCSE Chemistry** answer **all** of the questions in **Sections A and B**.
- Write your name and other details in the spaces provided above.
- Answer **all** questions in the spaces provided.
- Do all rough work on the paper.

Information for candidates

- The marks available are given in brackets at the end of each question or part-question.
- Marks will not be deducted for incorrect answers.
- In calculations show clearly how you work out your answers.
- There are **10** questions in **Section A** of this paper.
  There are **4** questions in **Section B** of this paper.
- There are no blank pages.

Advice to candidates

- Work steadily through the paper.
- Don't spend too long on one question.
- If you have time at the end, go back and check your answers.

1   Limestone is an important raw material. Some of its uses and products are shown in the diagram. Some are missing.

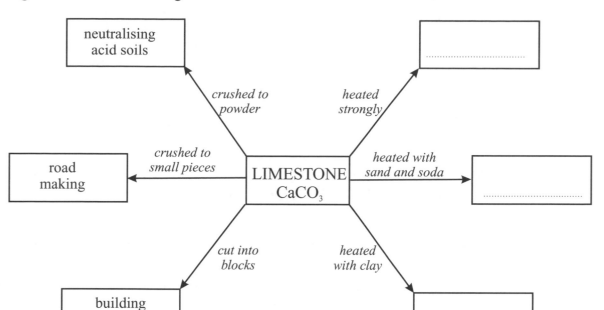

```
┌─────────────┐                              ┌─────────────┐
│ neutralising│                              │             │
│ acid soils  │                              │ ............│
└─────────────┘                              └─────────────┘
       ↖ crushed to              heated ↗
         powder                  strongly
┌─────────────┐  crushed to                   ┌─────────────┐
│   road      │  small pieces  ┌───────────┐  heated with  │             │
│  making     │ ←────────────  │ LIMESTONE │  sand and soda→│ ............│
└─────────────┘                │   CaCO₃   │               └─────────────┘
                               └───────────┘
       ↙ cut into            heated ↘
         blocks              with clay
┌─────────────┐                              ┌─────────────┐
│  building   │                              │   cement    │
│   stone     │                              │             │
└─────────────┘                              └─────────────┘
```

(a)   Write the names of the missing products in the spaces on the diagram.

(2 marks)

(b)   Bags of cement powder are labelled with this hazard symbol.

(i)   What does this hazard symbol tell you about cement powder?

....

(1 mark)

(ii)   Give **one** precaution you should take when handling cement powder.

....

....

(1 mark)

(c) The trout in a small lake were dying because the water was too acidic.
Limestone, $CaCO_3$, can be used to neutralise the acid.
The equation for this reaction is:

$$CaCO_3 \quad + \quad H_2SO_4 \quad \rightarrow \quad CaSO_4 \quad + \quad H_2O \quad + \quad CO_2$$

4900kg of sulphuric acid in the lake needed to be neutralised.
Calculate the mass of limestone, $CaCO_3$, which reacts with this amount of sulphuric acid.
You should show all your working.
(Relative atomic masses: H = 1, C = 12, O = 16, S = 32, Ca = 40)

....

....

....

....

....

....

(3 marks)

AQA, 2001

2   The table shows information about some compounds.

formula	name	type of structure	melting point in °C
CaO	calcium oxide	giant	2900
H_2O	water	molecular	0
NaCl	sodium chloride	giant	808
SO_2	sulphur dioxide	molecular	-75

(a) What links the melting point and the type of structure?
Use the information in the table to help you.

....

....

(2 marks)

QUESTION 2 CONTINUES ON THE NEXT PAGE

(b) Calcium and oxygen react together to form calcium oxide.

During the reaction two electrons move from a calcium atom to an oxygen atom.

Calcium ions, Ca^{2+}, and oxide ions, O^{2-}, are formed.

Finish the table. There are **two** spaces.

element	number of electrons in an atom	arrangement of electrons
calcium Ca	20	2.8.8.2
oxygen O	8	2.6

ion	number of electrons in an ion	arrangement of electrons
calcium Ca^{2+}	18	
oxide O^{2-}	10	

(2 marks)

(c) Calculate the relative formula mass of calcium oxide, CaO.

Use the Periodic Table to help you.

....

....

(2 marks)

(d) Strontium, Sr, reacts with oxygen in a similar way to calcium.

It forms a compound, strontium oxide, SrO.

Explain these facts.

Use your Periodic Table and your knowledge of the structure of atoms to help you.

....

....

....

....

(3 marks)

OCR, 2000

3   John Newlands attempted to classify the elements in 1866. He tried to arrange all the known elements in order of their atomic weights. The first 21 elements in Newlands' Table are shown below.

	COLUMN						
	a	b	c	d	e	f	g
Symbol	**H**	**Li**	**Be**	**B**	**C**	**N**	**O**
Atomic weight	1	2	3	4	5	6	7
Symbol	**F**	**Na**	**Mg**	**Al**	**Si**	**P**	**S**
Atomic weight	8	9	10	11	12	13	14
Symbol	**Cl**	**K**	**Ca**	**Cr**	**Ti**	**Mn**	**Fe**
Atomic weight	15	16	17	18	19	20	21

Use the Periodic Table to help you answer these questions.

(a) In two of Newlands' columns, the elements match the first three elements in two Groups of the modern Periodic Table.
Which two columns, **a** to **g**, are these?   .... and ....

(1 mark)

(b) (i)   A Group in the modern Periodic Table is completely missing from Newlands' Table. What is the number of this Group?
Group number ....

(1 mark)

(ii) Suggest a reason why this Group of elements is missing from Newlands' Table.

....

(1 mark)

(c) Give **one** difference between iron, Fe, and the other elements in column **g** of Newlands' Table

....

(1 mark)

(d) Give the name of the block of elements in the modern Periodic Table which contains Cr, Ti, Mn and Fe.

....

(1 mark)

QUESTION 3 CONTINUES ON THE NEXT PAGE

(e) Both Newlands and Mendeleev based their tables on atomic weights.
Explain why the modern Periodic Table is based on proton (atomic) numbers.

....

....

(2 marks)

(f) The atoms of elements in Group 1 of the modern Periodic Table increase in size going down the group.

Explain, in terms of electrons, how this increase in size affects the reactivity of these elements.

....

....

....

....

....

(3 marks)

AQA, 2000

4   The diagram shows stages in the manufacture of a fertiliser, ammonium sulphate.

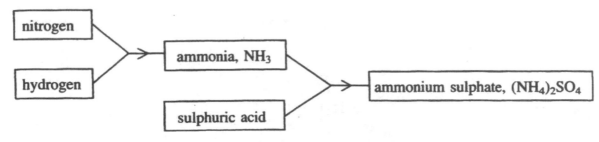

(a) Write a balanced equation for the formation of ammonia, indicating that the reaction is reversible.

....

(2 marks)

(b) Ammonia was first made by this reaction using a pressure of 25 atmospheres. Modern chemical plants use pressures of 200 atmospheres or more. State **two** advantages of using a higher pressure for this reaction. Give a reason to support each of your answers.

Advantage 1 ....

Reason ....

....

Advantage 2 ....

Reason ....

....

(4 marks)

(c) Ammonium sulphate is used to supply plants with increased levels of nitrogen. Calculate the relative formula mass of ammonium sulphate, $(NH_4)_2SO_4$, and hence the percentage by mass of nitrogen (N) in this compound.
(Relative atomic masses: H = 1.0; N = 14; O = 16; S = 32)

....

....

....

....

....

(3 marks)

Edexcel, 2001

5   The diagram represents the layers of the Earth

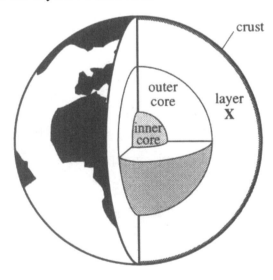

(a) (i)   Give the name of layer **X**

....

QUESTION 5 CONTINUES ON THE NEXT PAGE

(1 mark)

(ii) The inner core is smaller and nearer to the centre of the Earth than the outer core. Give **one** other difference between the inner and outer parts of the core.

....

(1 mark)

(b) A student was shown two igneous rocks A and B.
Rock A had large crystals. Rock B had small crystals.

(i) Describe how igneous rocks are formed.

....

....

....

(2 marks)

(ii) Explain fully why the crystals in rock A are larger than the crystals in rock B.

....

....

....

(1 mark)

(c) Explain why most igneous rocks form near the boundaries of tectonic plates.

....

....

....

....

....

....

(2 marks)

AQA, 2000

6   Octane is a saturated hydrocarbon.
     Its graphical (displayed) formula is

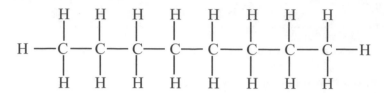

(a) Put a (ring) around the family to which octane belongs.

     alkanes   **alkenes**   **carbohydrates**   **carbonates**

(1 mark)

(b) The products formed when octane burns in air depend upon the amount of air.
     Explain this statement.

     ....

     ....

     ....

     ....

     ....

(4 marks)

(c) When octane vapour is passed over a heated catalyst a reaction takes place.
     Ethene and hydrogen are the only products.

     (i)   What type of reaction is taking place?

           ....

(1 mark)

     (ii) Finish and balance the symbol equation for this reaction.

           octane →   ethene   +   hydrogen

           C_8H_{18} → ....C_2H_4 + ...

(2 marks)

     (iii) Excess ethene is bubbled through bromine water.
           What colour change would you see?

           From .... to ....

(2 marks)

QUESTION 6 CONTINUES ON THE NEXT PAGE

Leave blank

(d) Ethene is used as the raw material for making poly(ethene).

    (i) Draw the graphical (displayed) formula of ethene and of poly(ethene).

ethene   poly(ethene)

(3 marks)

    (ii) Poly(ethene) has replaced paper and cardboard for many packaging uses. Suggest one advantage and one disadvantage of poly(ethene) compared to paper and cardboard. Do not consider the relative costs of the materials.

Advantage of poly(ethene) ....

....

Disadvantage of poly(ethene) ....

....

(2 marks)

OCR, 2000

7   Yoghurt is made by adding bacteria to milk. Enzymes in the bacteria change sugar in milk to lactic acid. This makes the milk go solid.

(a) Draw a ring around the best temperature to make yoghurt.

    0°C   **20°C**   **40°C**   **70°C**   **90°C**

(1mark)

(b) When yoghurt is made, how do you know that a chemical reaction has taken place?

....

(1 mark)

(c) What are enzymes and how do they affect chemical reactions?

....

....

....

(2 marks)

AQA, 2000

248

8   This question is about alkali metals and their compounds.

(a) An alkali metal (**X**) reacts violently with water. A gas (**Y**) and a solution (**Z**) are formed during this reaction. A lilac-coloured flame is seen.

(i) Name the substances **X**, **Y** and **Z**.

alkali metal **X** ....

gas **Y** ....

solution **Z** ....

(3 marks)

(ii) State what you SEE when universal indicator is added to solution **Z**. Give a reason for your answer.

....

....

(2 marks)

(b) Glauber's salt is a naturally occurring form of sodium sulphate. On heating, Glauber's salt loses water to form pure sodium sulphate. In an experiment, 20.0 g of Glauber's salt gives 8.8 g of pure, dry sodium sulphate when heated.

(i) What mass of water is lost?

....

(1 mark)

(ii) Calculate the percentage of water present in Glauber's salt.

....

....

(1 mark)

(iii) Give the names of the acid and the alkali which react together to produce sodium sulphate.

Acid ....

Alkali ....

(2 marks)

QUESTION 8 CONTINUES ON THE NEXT PAGE

(c) Find lithium (atomic number 3) in the periodic table.

   (i) Name a non-metal **in the same period** as lithium.

....

(1 mark)

   (ii) Name another metal **in the same period** as lithium

....

(1 mark)

(d) (i) Draw a diagram to show the arrangement of electrons in a lithium atom.

(1 mark)

   (ii) What is similar about the arrangement of electrons in the atoms of all the alkali metals?

....

....

....

(2 marks)

Edexel, 2000

9   The symbol equation shows the reaction between hydrogen and chlorine.

$$H_2 + Cl_2 \rightarrow 2HCl$$

BOND	ENERGY NEEDED TO BREAK BONDS OR RELEASED WHEN BONDS ARE FORMED (kJ per formula mass)
H—H	436
Cl—Cl	243
H—Cl	432

(a) When the formula masses of hydrogen and chlorine react, calculate:

   (i) the energy needed to break the bonds of the reactants.

....

....

....

(2 marks)

   (ii) the energy released when the bonds of the product are formed.

....

....

....

(1 mark)

(b) What do the figures in (i) and (ii) tell you about the overall reaction?

....

....
(2 marks)

(c) The reaction between hydrogen and chlorine is quite a fast reaction.
Other reactions are much slower but can be speeded up by using a catalyst.
Explain how a catalyst does this.

....

....
(1 mark)
AQA, 2000

10   This question is about the purification and uses of copper.

(a) The diagram shows the method of purifying a lump of impure copper by electrolysis.

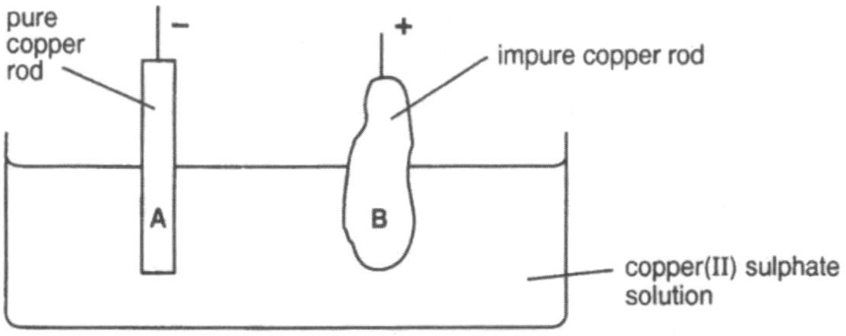

(i) What would you see happening at the electrodes during the electrolysis?

....
(1 mark)

(ii) What is the job of the copper(II) sulphate solution?

....
(1 mark)

(iii) Finish the ionic equations for the reactions at each electrode.

At **A** Cu^{2+} + ... $\rightarrow$ Cu

At **B** $Cu \rightarrow$ .... + $2e^-$
(2 marks)

(b) Copper is used for making electricity cables and water pipes in your home. What are the physical and chemical properties of copper which make it suitable for these uses?

electricity cables ....

....

water pipes...

....
(4 marks)
OCR specimen, 2003

11   Many soft drinks contain citric acid.

INGREDIENTS: CARBONATED WATER, SUGAR, CITRIC ACID, ACIDITY REGULATOR (E331), FLAVOURINGS, PRESERVATIVE (E211)

(a) Citric acid is a *weak acid*.

(i)   What is meant by a *weak acid* in terms of its ionisation in water?

....

....
(1 mark)

(ii) Describe and give the results of an experiment which would show that citric acid is a weaker acid than hydrochloric acid of the same concentration.

....

....

....
(2 marks)

(b) Citric acid behaves as an acid. Explain why, using the ideas of Arrhenius **and** of Bronsted-Lowry.

....

....

....
(3 marks)
AQA, 2003

12   Some students are given pieces of different metals.

They are asked to investigate which two metals would give the highest voltage when used in the cell. They use copper, zinc and magnesium

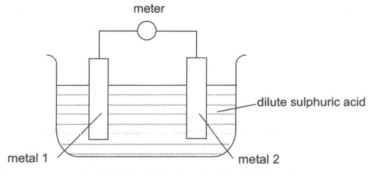

meter

dilute sulphuric acid

metal 1   metal 2

QUESTION 8 CONTINUES ON THE NEXT PAGE

Question 12 is only on the OCR syllabus

(a) (i) Which pair of metals gives the highest voltage?

metal 1 ....

metal 2 ....

(1 mark)

(ii) Give reasons for your choice.

....

....

(1 mark)

(b) What other change would affect the voltage of the cell?

....

(1 mark)

(c) In one cell zinc is used as metal 1 and copper is used as metal 2.

(i) Write an ionic equation for the reaction involving zinc in this cell.

....

(1 mark)

(ii) Describe and explain the flow of electrons in this cell.

....

....

....

(2 marks)

OCR, 2003 Specimen

13 (a) Describe, with the help of an ionic equation, what happens when silver nitrate, $AgNO_3$, solution is added to potassium bromide, KBr, solution.

....

....

....

....

....

(4 marks)

(b) Tom makes a solution of some iron(II) sulphate.

He tests this solution.

His results are shown in the table.

	test on iron(II) sulphate solution	result
Test 1	Add sodium hydroxide solution	A dark green precipitate **J** is formed.
Test 2	Bubble chlorine gas through, then add sodium hydroxide solution	A red-brown precipitate **L** is formed.

(i) Write down the name of precipitate **J**.

....

(1 mark)

(ii) Write down the name of precipitate **L**.

....

(1 mark)

(iii) Explain the function of the chlorine in the reaction with iron(II) sulphate.

....

....

....

(2 marks)

OCR, 2003 Specimen

14 A drain cleaner contains sodium hydroxide solution.

In a titration experiment, the sodium hydroxide in a 25.0 cm^3 sample of the drain cleaner was neutralised by 20.0 cm^3 of hydrochloric acid.

The concentration of the hydrochloric acid was 0.500 mol dm^{-3}.

The equation for the reaction is:

$$NaOH + HCl \rightarrow NaCl + H_2O$$

(a) Explain which bonds are broken, if any, and which are formed, if any, in the neutralisation reaction.

....

....

(2 marks)

254

(b) Describe, giving the names of the apparatus used, how the titration is carried out.

....

....

....

....

....

....

....

....

(4 marks)

(c) Calculate the concentration in mol dm^{-3} of sodium hydroxide in the drain cleaner.

....

....

....

....

....

....

(3 marks)

(d) Calculate the mass of sodium hydroxide in a bottle containing 250 cm^3 of this drain cleaner.

(Relative atomic masses: H = 1.0; O = 16; Na = 23)

....

....

....

....

....

....

(3 marks)

Edexcel, 2001

Page 4 (Warm-Up Questions)

1) Solid, liquid, gas.

2) Gases have a lot of free space between their molecules. Molecules in solids are already packed very closely together.

3) Heat energy makes some molecules move faster than others. Faster moving molecules at the surface overcome the forces of attraction between them and other molecules and escape.

4) As the water molecules in the ice warm up they vibrate more and more and eventually overcome the strong forces that hold them in a rigid lattice.

Page 5 (Exam Questions)

1 Atoms vibrate more (1) and the iron expands. (1)

2 It makes them move quicker, faster — they speed up. (1)

3 a) 3

   b) 4

   c) The heat energy supplied (put into the substance) (1) is used up in breaking bonds (the strong bonds between the particles in the solid) (1) rather than in raising the temperature. (1)

Page 12 (Warm-Up Questions)

1) Particle A = electron, Particle B = neutron, Particle C = proton

2) Atoms with the same atomic number but different mass numbers. *Or*, Atoms with the same number of protons but different numbers of neutrons. Carbon-14 has the same number of proton and electrons as Carbon-12, but has two more neutrons.

3)

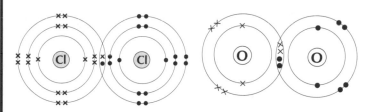

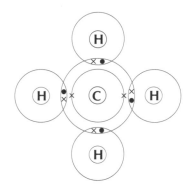

Page 13 (Exam Questions)

1 a) An atom (or group of atoms) that has lost (1) or gained electrons (1). This causes them to have a positive or negative charge.

   b) 10
      It's the atomic number (13) – 3. The mass number's just there to throw you.

2 a) i) 2,8

      ii) 2,8,8

   b) Ionic bonding.

   c)

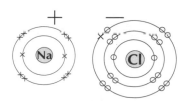

d)

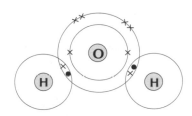

Page 21 (Warm-Up Questions)

1) Li^+, Na^+, K^+, Be^{2+}, Mg^{2+}, Ca^{2+}, O^{2-}, F^-, Cl^-

2) Water has a simple molecular structure with weak intermolecular forces. Diamond has a giant molecular structure with strong bonds holding all the atoms together.

3) Aluminium atoms are held together by metallic bonding (involving a sea of electrons) so some electrons are free to move and conduct electric current. When solid, the ions in salt can't move and so can't conduct electricity.

4) It turns limewater milky.

5) The contents are toxic.

Page 22 (Exam Questions)

1 a) Chlorine

   b) Oxygen

   c) Carbon dioxide

   d) e.g. Carbon dioxide, methane

2) a) Two from:
      Heat it to check that it has a boiling point of 100^0C.
      Add it to anhydrous copper sulphate. If this turns blue, the liquid is water. Check if it turns anhydrous cobalt chloride paper from blue to pink.

   b) Two from:
      Low melting point / low boiling point / liquid at room temperature / does not conduct electricity.

   c) Yes. Sodium chloride is ionic(1). When it dissolves, the ions are free to move through the solution(1). This means they can carry an electric current(1).
      Conducting electricity's always about those ions or electrons being able to move.

Page 23 (Exam Questions)

3 a) i) B

      ii) Metals have a giant structure with metallic bonding(1). This means that the outer electrons of each atom are free to move(1) through the metal, and can therefore carry an electric current(1).

   b) i) D

      ii) Bring a lighted splint near to it. If the gas ignites with a squeaky pop it is H_2.
      Ahh... the good old squeaky pop. Music to an examiner's ears.

   c) i) Ionic

      ii) Ionic lattices contain very strong chemical bonds between oppositely charged ions(1). This means it takes a lot of energy to break the bonds in the ionic lattice(1).

Page 29 (Warm-Up Questions)

1) Hydrogen and carbon

2) Fractional distillation

3) Carbon dioxide and water

4) Cracking

Page 30 (Exam Questions)

1 a) (i) A

      (ii) Kerosene(1). The more carbon atoms in the molecule, the higher the boiling range(1).

   b) Making plastics, surfacing roads, lubrication etc.

256

Page 31 (Exam Questions)

2 a) Carbon dioxide and water (1 mark for each)

   b) (i) Orange (1 mark) and smoky (1 mark)

      (ii) Ethane + oxygen → carbon dioxide + carbon monoxide + carbon + water (steam)
Balancing equations isn't covered until later, but if you're already able to do it, then try writing a balanced symbol equation for this reaction.

   c) Gas heaters can produce carbon monoxide by incomplete combustion of methane(1). This incomplete combustion happens when the heater is not sufficiently ventilated(1). Carbon monoxide is poisonous(1). It stops the blood from carrying oxygen. CO is odourless(1). (up to max 3 marks)

Page 36 (Warm-Up Questions)

1) Single covalent bonds

2) A hydrocarbon with no spare (double) bonds. An alkane.

3) Alkenes

4) No

5) Yes

Page 37 (Exam Questions)

1 a) Cracking hydrocarbons

   b) (i)

      (ii) Pressure (1) a catalyst (1)

      (iii) Polypropene is tough and elastic/not brittle(1) and forms durable fibres(1)

   c) (i) Polythene will not decompose, so it takes up space in landfill sites for a very long time.

      (ii) Save crude oil resources which are needed to make plastics.

      (iii) Consumers need to sort their rubbish or wash and reuse polythene products. Consumers might not want the hassle.

Page 42 (Warm-Up Questions)

1) A mineral containing enough metal to make the metal worth extracting.

2) Reduction with carbon in a blast furnace.

3) Aluminium.

4) As the metal.

Page 43 (Exam Questions)

1) a) (i) 1 mark for any from: Zinc, Iron, Tin, Lead (not Gold, as gold is found in the ground as a pure metal)
Look — these metals are all below carbon on the reactivity series. Easy marks.

      (ii) Electrolysis (1)

   b) (i) To make the coke burn faster so the temperature in the furnace increases.

      (ii) Silicon dioxide (SiO_2) (1) *or* sand

      (iii) Limestone is added to the blast furnace(1) and decomposes into calcium oxide and carbon dioxide(1). The calcium oxide reacts with silicon dioxide to make slag (1) which is then tapped off.
In order to get top marks here, you really need to know all those nasty little details. Go back and look at them again if they didn't all go in the first time.

Page 48 (Warm-Up Questions)

1) Aluminium

2) Aluminium

3) Bauxite

4) It's cheap and strong

Page 49 (Exam Questions)

1 a) For electrolysis to work, the Al^{3+} ions must be free to move(1). This is not possible in a solid (1).

   b) The melting point of aluminium oxide is over 2000°C (1). It would be very expensive to heat it to this temperature. Aluminium oxide dissolved in molten cryolite is liquid at a lower temperature which makes things cheaper (1).
When you're answering a question about an industrial process, remember that manufacturers are obsessed with cost. They want things done quickly using cheaper materials and cheaper processes. Well, they need to pay for their new Ferarris somehow.

   c) About 900°C

   d) (i) $Al^{3+} + 3e^- \rightarrow Al$ (1 mark for correct equation, 1 mark for balancing it)

      (ii) $2O^{2-} \rightarrow O_2 + 4e^-$ (1 mark for correct equation, 1 mark for balancing it)

      (iii) The graphite anode reacts with oxygen produced at the anode (1) to form CO_2(1).

Page 50 (Exam Questions)

2 a) It isn't pure enough to conduct electricity well.

   b) (i) A = the cathode
        B = the anode

      (ii) The cathode (A) gets bigger

3   Aluminium is higher in the reactivity series than carbon, and so can't be extracted by reduction with carbon (1). It needs to be extracted by electrolysis (1).

    Plus one more from:

    Electricity was not available for electrolysis before the late 1900s (1).
Electrolysis of aluminium ore was not discovered until the late 1900s (1).

Page 55 (Warm-Up Questions)

1) Clay

2) To neutralise unwanted acidity

3) A base

4) Iron

5) Eutrophication

Page 56 (Exam Questions)

1 a) Sand (silicon dioxide) and soda (sodium carbonate)

   b) $CaCO_3 \rightarrow CaO$ (1) $+ CO_2$ (1)

2 a) (i) High pressure (200-250 atmospheres) (1) and moderately high temperature (450°C)(1). Iron catalyst (1).

      (ii) At a lower temperature the reaction rate would be slower (1). Reducing the speed would make production less profitable (1).

   b) (i) $NH_{3(g)} + HNO_{3(aq)} \rightarrow NH_4NO_{3(aq)}$ (1 mark for correct equation, 1 mark for balancing it).

      (ii) The fertilisers promote excessive growth of plants and algae(1). Then the plants die off and bacteria feed off them, using up the oxygen in the water (1). Fish and more plants die because there isn't enough oxygen in the water(1).

Page 62 (Warm-Up Questions)

1) Endothermic

2) Neutralisation

3) Precipitation

4) Negative

5) $2H^+ + 2e^- \rightarrow H_2$

Page 63 (Exam Questions)

1 a) No

   b) Hydrogen (1). Sodium is too reactive to form (1).

   c) $2Br^-_{(aq)} \rightarrow Br_{2(g)} + 2e^-$.
The positive electrode = anode.

2 a) Oxygen relights a glowing splint

   b) At the positive electrode: $2H_2O_{(l)} \rightarrow O_{2(g)} + 4H^+_{(aq)} + 4e^-$
    At the negative electrode: $2Cu^{2+}_{(aq)} + 4e^- \rightarrow 2Cu_{(s)}$

THE ANSWERS

Page 67 (Warm-Up Questions)

1) 4
2) beryllium 5, sodium 12, magnesium 12.
3) Na_2CO_3 = 106, $(NH_4)_2SO_4$ = 132, $Ca(OH)_2$ = 74
4) NH_3 = 82.4%, NH_4Cl =26.2%, NH_4NO_3 = 35%

Page 68 (Exam Questions)

1 (a) carbon = 40g, hydrogen = 6.7g, oxygen = 53.3g
   (b) C:H:O = 1:2:1 Empirical formula : CH_2O (1 mark for ratios, 1 for formula)
2 (a) Cu = 40%, S = 20%, O = 40%
   (b) 72g
3 60%

Page 73 (Warm-Up Questions)

1) 137.5g
2) 2,625 cm^3 or 2.63 litres
3) a) 127g
   b) 77g
   c) About 26,100 cm^3 or 26.1 litres (dm^3)
4) 10g

Page 75 (Exam Questions)

1 20.2
2 a) Fe_2O_3 + 3CO → 2Fe + $3CO_2$
   b) 56g
   c) 36 litres (dm^3) or 36,000cm^3
   d) Any two from: making steel, construction, cars, lorries, trains, boats etc (not planes)

Page 76 (Exam Questions)

3 a) N_2 + $3H_2$ ⇌ $2NH_3$
   b) 180 litres (dm^3) or 180,000cm^3
4 a) $4Ag^+$ + $4e^-$ → 4Ag
      $2O^{2-}$ → $4e^-$ + O_2
   b) 1.1g
   c) 1528cm^3 or approximately 1.5 litres (dm^3)

Page 79 (Warm-Up Questions)

1) 6.023×10^{23}
2) 19g
3) 80g
4) 80g
5) 42g

Page 80 (Exam Questions)

1 a) i) 84g
      ii) 1 mole
      iii) 24 litres
   b) 1.004×10^{23}
   c) 2.5 litres

Page 87 (Warm-Up Questions)

1) Main gases: Carbon dioxide, carbon monoxide — also ammonia and methane. Also accept steam or water vapour.
2) Methane and ammonia reacted with oxygen to release nitrogen Nitrogen was released by denitrifying bacteria during the decay of organic matter.
3) Argon
4) Photosynthesis. Dissolving in sea/rain water
5) Lakes/soils become acidic. Plants/trees/fish die. Limestone/marble buildings are damaged
6) O_3

Page 88 (Exam Questions)

1 1= photosynthesis
   2= respiration
   3= decay (or burning)
   4= burning
2 a) Oxygen increased as green plants photosynthesised.
   b) Carbon dioxide decreased as water vapour condensed to form oceans, and rain, which dissolved it.

Page 89 (Exam Questions)

3 a) 21cm^3
   b) 21%
   c)

Gas	Percentage
Nitrogen	78%
Carbon dioxide	0.04%
Argon	1%
Oxygen	21%

We know these add up to 100.04% — we've just rounded them a little. The figures you get in the exam may be very slightly different, so be careful.

4 a) Sulphur dioxide and nitrogen oxides
   b) Burning fossil fuels

Page 96 (Warm-Up Questions)

1) Limestone, sedimentary
2) The clay gets heated and compressed
3) Magma
4) e.g. Sandstone – buildings
   Marble – headstones, statues
5) Both formed from molten magma that cools and solidifies. Intrusive cool slowly underground and extrusive cool quickly on the surface.

Page 97 (Exam Questions)

1 a) i) C or D or E
      ii) A
      iii) B or F
   b) i) A
      ii) F
      iii) D or E
   c) Rock B may have cooled slowly underground, so that bigger crystals formed than in rock F, which may have cooled quickly on the surface. Both are igneous rocks, but rock B is intrusive, rock F is extrusive.

Page 98 (Exam Questions)

2 i) the sea, lakes etc.
   ii) sedimentary rocks
   iii) heat and pressure
   iv) metamorphic rocks
3 a) Over millions of years, eroded rock particles are transported to and deposited in the sea(1); the particles form layers that are gently squashed together(1); the water is squeezed out and salts/minerals cement the particles together(1)
   b) Like concrete – pebbles stuck together by finer particles
   c) metamorphic
4 a) P = sedimentary
      Q = igneous
   b) Rock Q is formed as molten magma cools and solidifies inside the earth.

Page 102 (Warm-Up Questions)

1) Water gets into cracks in rock, freezes and expands. The cracks widen until rock fragments break off

2) Biological weathering, chemical weathering

3) Erosion is the wearing away of rocks by any means.

4) Any one from: volcano, large cracks, mountains, ocean trench

5) The collision between an oceanic plate and a continental plate forced the oceanic plate underneath. It melted, leading to magma at pressure below. The magma pushed up through the crust to form volcanos. The continental crust crumpled to form mountains.

Page 103 (Exam Questions)

1 a) (Tectonic) plates

  b) i) X may be a continental plate, while Y is an oceanic plate, rock X is less dense than rock Y

    ii) It will start to melt and turn into magma

2 a) Any two from:
    Physical weathering: water in cracks in rock freezes, expands, and breaks the rock apart
    Chemical weathering: acidity in rain dissolves limestone and marble
    Biological weathering: the growth of plant roots in cracks in rock breaks the rock apart

  b) the sun

Page 110 (Warm-Up Questions)

1) False

2) 2

3) 1

4) aluminium

Page 111 (Exam Questions)

1 a) S, sulphur

  b) No

2 a) A and B

  b) C

  c) A

Page 116 (Warm-Up Questions)

1) Li^+

2) No

3) Noble gases have full outer electron shells — they do not form any bonds. Oxygen and fluorine have incomplete outer shells and do form bonds.

4) Low

5) Lithium, sodium or potassium

Page 117 (Exam Questions)

1 a) Potassium hydroxide and hydrogen

  b) $2K + 2H_2O \rightarrow 2KOH + H_2$

  c) Sodium has fewer electron shells than potassium(1) so the outer electron is closer to and less shielded from the nucleus(1). This means sodium gives up its outer electron less readily(1).
    I can't stress how vitally important it is that you learn about shielding. It's really, really important for understanding why reactivity changes down the groups.

2 a) i) They have complete outer electron shells

    ii) A monatomic gas

  b) i) Neon

    ii) It gives an inert atmosphere which stops the filament from burning away

  c) Helium is inert. It won't burn or explode. Also it's very light.

Page 124 (Warm-Up Questions)

1) yellow

2) liquid

3) sodium chloride

4) because chlorine kills germs — makes the swimming pool safer.

5) because it helps prevent tooth decay.

Page 125 (Exam Questions)

1 a) Adding salt lowers the freezing point of water, and melts the ice. The grit in the rock salt adds extra friction.

  b) Sea water is placed in big open tanks. The sun evaporates off the water, leaving the salt behind.
    The reference to 'hot countries' is a dead give away — it's got to be to do with the sun and evaporation.

  c) i) $2AgBr \rightarrow Br_2 + 2Ag$

    ii) black

2 a) It is bonded covalently

  b) i) $HCl_{(g)} \xrightarrow{water} H^+_{(aq)} + Cl^-_{(aq)}$

    ii) The H^+ ions.

Page 126 (Exam Questions)

3 a) Elements at the top of Group VII have fewer electron shells(1) so the attraction of the +ve nucleus is less shielded(1). This means they attract electrons from other atoms more strongly.

  b) i) $Cl_2 + 2KI \rightarrow I_2 + 2KCl$

    ii) No. Iodine will not react with potassium chloride because iodine is lower down Group VII (1) so it is less reactive than chlorine (1)

4 a) Sodium chloride

  b) (answers can be in any order)

    i) Hydrogen (1). 1 mark for any use from: Haber Process to make ammonia, hydrogenating oil to make margarine

    ii) Chlorine (1). 1 mark for any use from: Making bleach, making hydrochloric acid, making insecticide, making plastics (eg PVC)

    iii) Sodium hydroxide (1 mark). 1 mark for any use from: Making soap, making various organic chemicals, oven cleaner, making paper, ceramics manufacture

Page 133 (Warm-Up Questions)

1) red

2) alkali

3) alkali

4) H^+

5) water

6) water + carbon dioxide

Page 135 (Exam Questions)

1 a) Alkaline

  b) i) $2HNO_3 + MgO \rightarrow Mg(NO_3)_2 + H_2O$

    ii) $2HCl + CaO \rightarrow CaCl_2 + H_2O$

    iii) Do not react, as both compounds are alkaline.

    iv) $H_2SO_4 + CuO \rightarrow CuSO_4 + H_2O$

  c) i) Reddish orange (weak acid)

    ii) Red (strong acid)

    iii) Sulphur oxides dissolve to form sulphuric acid. Nitrogen oxides dissolve to form nitric acid(1). These fall as acid rain(1), which corrodes buildings and statues, and kills trees and fish(1).

Page 136 (Exam Questions)

2 a) i) Calcium carbonate

    ii) $CaCO_3 + 2HCl \rightarrow CaCl_2 + H_2O + CO_2$ (1 mark for right reactants, 1 for right products, and 1 for balancing it)

    iii) The limewater will turn milky

  b) $NaHCO_3 + HCl \rightarrow NaCl + H_2O + CO_2$

3 a) Ammonium chloride NH_4Cl

  b) i) Fertiliser

    ii) It has more nitrogen, which plants need.

Page 140 (Warm-Up Questions)

1) non-metal
2) non-metal
3) copper and tin
4) metals
5) no

Page 141 (Exam Questions)

1 a) i) The shared outer electrons move freely and carry heat energy through the structure

   ii) If the attractive force between atoms is strong, it will take a lot of energy to break it. This gives metals high melting points.

 b) i) Oxygen is a diatomic molecule. Two atoms of oxygen form a covalent bond by sharing electrons

   ii) Diamond has a giant tetrahedral crystal structure. Covalent bonds are formed between one carbon atom and four other carbon atoms around it.

 c) i) Graphite

   ii) Between the two layers of crystal there are free electrons (because each carbon atom has four electrons in its outer shell, but has only made three bonds). These electrons move freely which means that graphite conducts electricity.

Page 146 (Warm-Up Questions)

1) yes
2) no. It will react with steam.
3) aluminium
4) yes
5) zinc and copper
6) yes

Page 147 (Exam Questions)

1 a) Gold, silver and sometimes copper

 b) Potassium, sodium, calcium, magnesium, aluminium

 c) Silver is lower down the series than hydrogen. Only metals above hydrogen in the reactivity series react with water.

2 a) Yes, as they are typical metals with a giant metallic structure

 b) Nickel

 c) i) Cu^+

   ii) Purple   *See, you really do need all those fiddly details.*

Page 155 (Warm-Up Questions)

1) temperature, concentration (or pressure), catalyst, surface area
2) Increase the temperature
3) Three from: decrease the temperature, decrease the concentration, decrease the pressure (for a gas), decrease the surface area (or use bigger pieces of solid).

   Remember — there are lots of ways to increase the number of collisions, but raising the temperature's the only way to make the collisions faster.

4) Two from: manganese (IV) oxide, liver, blood, celery, potatoes, catalase, peroxidase.
5) Place the beaker on some scales. Record the fall in mass at regular intervals. Attach a gas syringe to the beaker with an air-tight seal. Record the volume of gas given off at regular intervals.

Page 157 (Exam Questions)

1 a)

b)

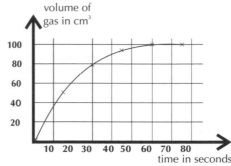

i) 2 marks if all points correctly plotted; 1 mark if 4 or 5 points correctly plotted.

Graph questions like this are easy marks — but take care not to slip up when the exam pressure's on.

ii) 1 mark for line of best fit drawn as above.

2 a) The mass will decrease

 b) 7 minutes

 c) i)

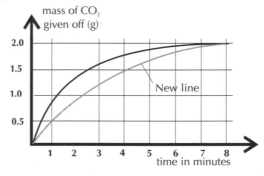

   ii) For a certain mass of marble chips, the total surface area of bigger pieces is less than the total surface area of smaller pieces (1). The smaller the surface area, the smaller the number of exposed particles (1) and the fewer successful collisions (1)

Page 165 (Warm-Up Questions)

1) A biological catalyst — it speeds up reactions in living things without the need for high temperatures, and is not used up.
2) About 45°C
3) fermentation
4) They can be mixed into plastic beads, or trapped in an alginate bed.
5) Proteases digest proteins. Lipases digest fats.

Page 166 (Exam Questions)

1 a) i) Fermentation is the process of yeast converting sugar(1) to carbon dioxide and alcohol(1).

   ii) Glucose $\xrightarrow{zymase}$ carbon dioxide + ethanol (1 mark for reactants, 1 mark for products)

   iii) Too hot and the enzyme is destroyed. Too cold and the enzyme will work very slowly.

 b) i) No — the enzymes in the detergent work best at 40°C and would be destroyed at high temperatures.

   ii) Protease

   Enzymes = biological catalysts. They're pretty fussy and work best at a certain temperature and pH.

2 a) Enzymes are used to turn the starch in corn into sugar.

 b) You need less fructose to give the same sweetness as sucrose.

 c) Zymase

Page 167 (Exam Questions)

3 a) i) Iron

    ii) Platinum

  b) Enzymes are not scarce(1). They are more specific(1) than inorganic catalysts. They work at low temperatures and pressures(1), so reaction vessels don't need to be heated or kept at high pressure, which is expensive(1).

    Yet again, it all comes back to saving money.

4 a) The enzyme starts to be destroyed when the temperature reaches 38°C.

  b) i) Bacterial enzymes make food go off(1). Freezing virtually stops enzyme activity(1), so the bacterial enzymes can't make the food go off.

    ii) Yes(1). Once the food is thawed, the bacterial enzymes will work again(1).

Page 174 (Warm-Up Questions)

1) Yes

2) The water is driven off to leave white anhydrous copper (II) sulphate.

3) A system from which no reactants or products can escape.

4) No

Page 175 (Exam Questions)

1 a) In dynamic equilibrium, the forward and reverse reactions are going at exactly the same rate, with no net change in % of products or reactants.

  b) i) endothermic

    ii) The reaction AB $\rightarrow$ A + B will speed up and the reaction A + B $\rightarrow$ AB will slow down.

    iii) It will decrease.

  c) If you change the conditions of a reversible reaction, the position of equilibrium will shift to oppose the change.

    This is something you've just got to know. That's all there is to it.

2 a) A reversible reaction is one where the products of the reaction can themselves react to produce the original reactants.

  b) The reaction has not stopped. The forward and backwards reactions are occurring at the same rate.

Page 176 (Exam Questions)

3 a) i) Higher pressure favours the forward reaction (1) because there are four volumes of gas on the reactant side and only two volumes of gas on the product side (1). Increasing pressure will favour the side with least volume.

    ii) At lower temperatures the rate of reaction would be too slow (1). It's not worth waiting a lot longer for a slightly better yield (1). 450°C is a compromise between rate of reaction and yield. (1)

    There's no perfect solution that'll give the maximum amount of ammonia really quickly. Manufacturers just have to <u>compromise</u>.

    iii) The yield would increase(1). The concentration of ammonia would be reduced(1), and so the rate of the forward reaction would increase to give more ammonia(1).

  b) i) It has no effect on the yield.

    ii) High temperatures favour the backward reaction, and reduce the yield(1). The catalyst speeds up the reaction without the temperature being raised and the yield dropping(1).

Page 181 (Warm-Up Questions)

1) Exothermic

2) Endothermic

3) Endothermic

4) Takes in energy

Page 182 (Exam Questions)

1 a) 1 mark for any example from: burning fuels, neutralisation reactions (or any specific reaction between an acid and an alkali), addition of water to anhydrous copper(II) sulphate, etc.

  b) i) ΔH is negative

    ii)

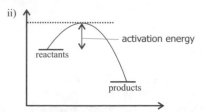

  c) Making new bonds is exothermic

Page 183 (Exam Questions)

2 a) Activation energy

  b)

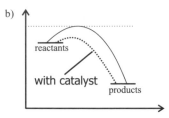

  c) The overall energy change stays the same.

3    $C_2H_4 + 3O_2 \rightarrow 2CO_2 + 2H_2O_{(g)}$

    Bonds broken: Three O=O bonds in oxygen, four C-H single bonds and one C=C double bond in ethene(1). (OR 1 mark for writing down (3×498) + (4×414) + 615)

    (4×414) + 615 + (3×498) = 3765 kJ/mol(1)

    Bonds formed: Four C=O double bonds and four H-O single bonds. (1) (OR 1 mark for writing down (4×749) + (4×463)

    (4×749) + (4×463) = 4848 kJ/mol(1)

    Difference = 1083 kJ/mole (1 mark for final answer)

    Bond energy calculations are simple — you just need to practise them. Remember, it's vitally important to check how many moles of each thing there are.

Page 192 (Warm-Up Questions)

1) acids

2) bases

3) silver chloride and lead chloride

4) barium sulphate, lead sulphate. (calcium sulphate is only very slightly soluble)

5) SO_4^{2-} (sulphate)

6) I^- (iodide)

  There's no trick for getting this stuff right. It's up to you to learn all the tests off by heart — you don't know which they'll throw at you in the exam.

Page 193 (Exam Questions)

1 a) i) $Ca(NO_3)_2 + H_2SO_4 \rightarrow CaSO_4 + 2HNO_3$

    ii) No more calcium sulphate precipitates out

  b) Add magnesium metal or insoluble magnesium hydroxide or magnesium carbonate (1) to a beaker of sulphuric acid (1). The reaction is finished when no more bubbles of gas are produced, and excess metal/hydroxide/carbonate remains in the beaker (1). Filter off the excess solid, and evaporate off the water to leave solid magnesium sulphate(1).

  c) The hydroxides and carbonates of these three substances are soluble. Excess will dissolve, so you can't add excess and then filter it off(1). Titration allows you to add just the right amount of each reactant so there is no excess(1).

2 a) Brønsted and Lowry said that acids released H^+ ions (are proton donors) (1) and bases took up H^+ ions (are proton acceptors) (1).

b) $H_2SO_4 \rightarrow 2H^+ + SO_4^{2-}$

c) Strong acids ionise completely in aqueous solution. Weak acids only partially ionise.
   Don't fall into the trap of confusing weak acids with dilute acids. 'Dilute' just means it's watered down.

Page 194 (Exam Questions)

3 a) i) Carbon dioxide

   ii) Gases are more soluble under high pressure. When the cap comes off, the pressure goes down and the bubbles come out of solution. (1 mark for suggesting that the pressure goes down. 1 mark for suggesting that the carbon dioxide is more soluble at high pressure than at low pressure)

b) The liquid is put in a side-arm flask(1). A gas syringe is attached to the flask(1). The solid is added, and the flask quickly sealed. The amount of gas evolved is measured in the gas syringe(1).

c) Ammonia gas is very soluble(1), and will dissolve to form ammonia solution instead of passing through the water and collecting in the collection vessel(1).

4 a)

ion	reagent(s) added	colour of precipitate
Fe^{2+}	NaOH	dark green
Ca^{2+}	NaOH	white
SO_4^{2-}	HCl + BaCl$_2$	white

b) $AlBr_3$

Page 198 (Warm-Up Questions)

1) pink

2) Swirling ensures that the solutions are properly mixed. Going slowly ensures that you don't accidentally add too much.

3) flame tests

4) Infra-red spectroscopy

Page 199 (Exam Questions)

1 a) Moles of acid used = 25/1000 × 0.2 = 0.005

   Moles of Na_2CO_3 = half moles of acid = 0.0025

   Moles in 1 dm³ = (0.0025 × 1000)/50 = 0.05

   So there are 0.05 moles Na_2CO_3 in 14.3g of the hydrate
   Don't forget that 1dm³ = 1000 cm³, or you'll find yourself in a right muddle.

b) 0.05 moles in 14.3g of the hydrate.

   1 mole in (14.3×20) of hydrate = 286g

   1 mole anhydrous Na_2CO_3 = 23+23+12+16+16+16 = 106g

   Mass of water in 1 mole = 286-106=180.

   RMM of water = 18g.

   180/18 = 10 moles of water

   Formula = $Na_2CO_3.10H_2O$
   This sort of question calls for clear thinking. To get all the marks you can, be sure to show each step of your working.

Page 200 (Exam Questions)

2 Measure hydroxide in pipette. (1 mark for hydroxide in pipette and acid in burette, or the other way round)

   Add a few drops of indicator. (1)

   Add acid/hydroxide slowly from burette. Swirl the flask as you do. (1)

   Titrate until endpoint (indicator changes colour) (1)

   Add 1 drop at a time towards the end. Note down the volume of hydroxide needed.

   Repeat without indicator (1)

3 a) 20cm³

b) To start with there are many OH⁻ ions. The number of ions decreases as the reaction proceeds towards the endpoint(1). Then the number of H⁺ ions increases after the endpoint(1). The more ions are present, the higher the conductivity(1).

Page 204 (Warm-Up Questions)

1) carbonic acid, H_2CO_3

2) Ca^{2+}, Mg^{2+}

3) Any three from: bacteria, colloidal particles, acidity, nasty smells/odours.

4) Very small gas bubbles dispersed in a liquid.

5) Replacing calcium and magnesium ions in water with sodium or hydrogen ions — removing hardness.

Page 205 (Exam Questions)

1 a) A colloid is tiny particles of one substance dispersed in another

b) The particles are negatively charged. Water molecules are slightly polar. The positive end of a water molecule is attracted to the negatively charged particles. The water molecules around the charged particle have their negative ends on the outside, so they stay apart by mutual repulsion.

c) The metal ion is attracted to the negatively charged colloid particles. They bond to form neutral molecules which precipitate out.

2 a) To kill micro-organisms/bacteria

b) By passing the water through a carbon slurry

c) i) $CO_2 + H_2O \rightarrow H_2CO_3$

   ii) By passing the water through a limestone slurry.

   $2H^+ + CaCO_3 \rightarrow Ca^{2+} + CO_2 + H_2O$
   This last equation isn't totally new — you know an acid plus a carbonate gives a salt, carbon dioxide and water. See page 131 if you don't.

Page 209 (Warm-Up Questions)

1) 600 coulombs

2) 96000 coulombs, 1 mole

3) 2

4) anode

5) Oxidation Is Loss, Reduction Is Gain (of electrons)

6) Magnesium and silver

Page 210 (Exam Questions)

1 a) i) Hydrogen

   ii) Chlorine

   iii) One mole of chlorine is produced for every mole of hydrogen ($2H^+ + 2e^- \rightarrow H_2$ and $2Cl^- \rightarrow Cl_2 + 2e^-$). 1 mole of any gas takes up the same volume.

2 a) Magnesium and copper. Magnesium is the most reactive and copper is the least reactive. The bigger the difference in reactivity, the bigger the voltage.

b) The solution turns blue — copper electrode dissolves to form blue Cu^{2+} ions

c) i) 0.76 + 0.8 = 1.56V

   ii) 0.13 + 0.34 = 0.47V

3 a) $H_2 \rightarrow 2H^+ + 2e^-$

b) $O_2 + 4e^- \rightarrow 2O^{2-}$

Page 216 (Warm-Up Questions)

1) pure aluminium is too soft

2) nuclear reactors, jet engines and replacement joints, for example

3) the main ore of titanium, TiO_2

4) no, it's heavy and brittle

5) 0.25%

6) cutting tools, for example

Page 217 (Exam Questions)

1 a) Aluminium forms a layer of oxide on its surface which protects it against further corrosion

b) i) Sulphuric acid

   ii) $4OH^- \rightarrow 2H_2O + O_2 + 4e^-$

2 a) Strong, light (low density), corrosion resistant

b) This process is very expensive, so titanium is only used when it's really important and when it can be afforded.

3 a) Oxygen is blasted through. This reacts with carbon impurities to form CO_2 which comes off as gas

   b) Chromium and nickel

   c) Galvanising is coating the steel with a thin covering of zinc.
      Questions on the nitty gritty of steel production do have a nasty habit of cropping up. It's not exactly riveting stuff, but you need to know it anyway.

Page 220 (Warm-Up Questions)

1) SO_2

2) $2SO_2 + O_2 \rightleftharpoons 2SO_3$

3) 450°C, 1-2 atmospheres, vanadium pentoxide catalyst

4) Under strong UV light, e.g. in the upper atmosphere.

5) Damage to the ozone layer means that more UV light gets through the atmosphere. UV light is harmful to living things.

Page 221 (Exam Questions)

1 a) i) High pressure

       ii) High pressure would liquify the SO_2.

      iii) Low temperature

       iv) The rate of reaction would be too slow

   b) It increases the rate of reaction by lowering the activation energy

2 a) Another free radical is generated in the reaction, which can go on and destroy another ozone molecule and so on and so on

   b) The reaction needs ultraviolet light. Levels of UV light are high enough in the upper atmosphere, and not at ground level

   c) They are concerned about further damage to the ozone layer, which could lead to health problems like skin cancer.

Page 227 (Warm-Up Questions)

1) isomers

2) The more chain branching, the lower the boiling point.

3) to make it undrinkable

4) weak
   Don't be fazed by carboxylic acids — they do normal acid things. Make sure you can name them, draw them and know how they're used though.

5) butyl methanoate

Page 228 (Exam Questions)

1 a) $2CH_3OH + 3O_2 \rightarrow 2CO_2 + 4H_2O$

   b) It burns to produce only CO_2 and H_2O, doesn't smell and washes off easily unlike petrol/oil. Can be cheaply produced from crops like sugar cane.

   c) i) $2C_2H_5OH + 2Na \rightarrow 2C_2H_5ONa + H_2$

       ii) Reaction of sodium with water

2 a) i) $C_6H_{12}O_6 \rightarrow \mathbf{2C_2H_5OH} + 2CO_2$

       ii) Oxygen will oxidise ethanol to ethanoic acid (1 mark for "make it sour" or "make it go off"). Also, oxygen allows the yeast to respire aerobically, and therefore not produce alcohol.

   b) i) Ethene and water

       ii) 300°C, 70 atmospheres pressure, phosphoric acid catalyst

Page 229 (Exam Questions)

3 a) Vinegar

   b) Ethanol + oxygen → ethanoic acid

   c) Citric acid

   d) 1 mark for any from apple, pear, fruit, rum etc.

4 a) esters and water

   b) i) Ethyl methanoate
           This is really easy if you know that the alcohol (ethanol, in this case) gives the first part of the name, and the acid (methanoic acid) gives the second part — learn it.

       ii) An acid, e.g. sulphuric acid

      iii) A fruity/pleasant smell

c)

$$H-C\overset{O}{\underset{O}{\diagup}} \quad \overset{H}{\underset{H}{\overset{|}{C}}} - \overset{H}{\underset{H}{\overset{|}{C}}} - H \quad + \quad H_2O$$

Page 234 (Warm-Up Questions)

1) fats, carbohydrates, proteins

2) $C_6H_{12}O_6$

3) amino acids

4) yeast or baking soda

5) fats

6) aspirin

Page 235 (Exam Questions)

1 a) i) The COOH part

       ii) William is right. A molecule of water is formed for each amino acid that joins on. Any polymerisation that forms water like this is a condensation polymerisation.

   b) i) Proteins (accept polypeptides)
           The amino acids are joined together by peptide links — that's why they're called polypeptides.

       ii) Hydrolysis

2 a) Iron, needed to make haemoglobin (1) OR Calcium, needed for strong teeth and bones. (or any other suitable answer)

   b) Vitamin C

Page 236 (Exam Questions)

3 a) i)

$$\begin{array}{c} H \\ | \\ H-C-OH \\ | \\ H-C-OH \\ | \\ H-C-OH \\ | \\ H \end{array}$$

       ii) Fatty acids/carboxylic acids

   b) i) Saturated fats are made with fatty acids that have no double bonds. Unsaturated fats are made with fatty acids that have at least one double bond between carbon atoms.

       ii) Unsaturated fat

4 a) Aspirin thins the blood, reducing the risk of heart attack

   b) i) To minimise the risk to human test subjects / if the drug is harmful, it will only damage tissue culture or rats.

       ii) A placebo is a pill that looks just like the drug being tested, but has no active ingredients at all. Sometimes medical conditions can improve just because the patient thinks they are being treated. By comparing patients given the real drug to those given a placebo this variable can be accounted for.

THE ANSWERS

Exam Paper

Please note: The answers to the past exam questions have not been provided by or approved by the examining bodies (AQA, OCR and London Qualifications Ltd - Edexcel). As such, AQA, OCR and London Qualifications Ltd do not accept any responsibility for the accuracy or method of the working in the answers given. CGP has provided suggested solutions — other possible solutions may be equally correct.

SECTION A

1 a) Heated strongly limestone turns to calcium oxide (quicklime) *(1 mark)*
Heated with sand and soda, limestone turns to glass *(1 mark)*

   b) i) This label on a substance means that it is an irritant. *(1 mark)*

      ii) Put on gloves before handling *(1 mark)*.

   c) M_r of $H_2SO_4 \Rightarrow (2 \times 1) + (1 \times 32) + (4 \times 16) = 98$
   4 900 000 / 98 = 50 000 moles

   M_r of $CaCO_3 \Rightarrow (1 \times 40) + (1 \times 12) + (3 \times 16) = 100$
   1 mole of $CaCO_3$ weighs 100 g, so 50 000 moles weigh 5 000 000 g, or 5 000 kg. *(3 marks)*

2 a) Giant structures have high melting points *(1 mark)*. Molecular structures have low melting points / melt easily *(1 mark)*. ("Molecular structures have lower melting points than giant structures" gains 2 marks)

   b) 2.8.8 *(1 mark)* (accept 2 8 8 or 2-8-8). 2.8 *(1 mark)* (accept correct diagrams)
   Don't forget — an atom's main aim in life is to end up with a full shell of electrons.

   c) 40 + 16 *(1 mark)*
   = 56 *(1 mark)* (award 1 mark for 20 + 8 = 28)

   d) Both in the same Group (or words to that effect) / they have the same outer electrons **or** Strontium has two outer electrons / strontium is in Group 2 *(1 mark)*. Strontium loses two electrons / forms an ion Sr^{2+} *(1 mark)* (Reject "electrons shared")

3 a) b and c (both required, accept upper case) *(1 mark)*.

   b) i) Group 0 / Group 8 / Group 18 (only accept these **numbers**) or noble gases / inert gases *(1 mark)*

      ii) Not discovered etc. / could talk about the lack of methods to isolate them (do **not** accept 'not made then') *(1 mark)*

   c) e.g. iron is a metal / the others are non-metals / any valid physical or chemical difference such as iron conducts electricity (reject atomic weight / mass). (Do **not** accept "iron is a solid" because sulphur is also solid) *(1 mark)*

   d) Transition (metals) / transitional / d-block *(1 mark)*
   Pretty much all you need to do in this question so far is find your way around the Periodic Table. That's not too hard.

   e) for **two marks**:
   proton number = number of electrons

   properties/reactions of elements depend on number/arrangement of electrons / elements with similar properties go in the same group / elements in the same group have the same number of electrons in outer shell
   OR
   for **two marks**:
   elements with similar properties go in the same group

   when elements are placed in order of relative atomic mass they do not all go in the correct groups / or example
   OR
   for **one mark**:
   proton number determines the properties/reactions

   proton number defines the element / each element has a different proton number the relative atomic mass is an average of the masses of the isotopes of an element / different elements can have the same atomic mass.

   f) Reactivity increases down the group

   Outer / valence electrons more easily lost

   Force of attraction on outer / valence electrons decreases / more shielding / further from pull of nucleus etc. (accept converse argument)
   (if there is no reference to outer shell / valence then limit to two marks)
   (1 mark for each up to a total of 3 marks)

4 a) $N_2 + 3H_2 \rightleftharpoons 2NH_3$, balanced correct formula *(1 mark)* and reversible arrow *(1 mark)*

   b) Advantage 1: faster reaction
   Reason: more (frequent) collisions / molecules pushed closer together
   Advantage 2: higher yield
   Reason: fewer molecules on RHS / left to right reaction involves decrease in volume / equilibrium moves to right (Le Chatelier's Principle) *(4 marks)*

   c) $(NH_4)_2SO_4$ = 132

   Percentage of N = $100 \times \dfrac{28}{132}$

   $= 21(.2)$ *(3 marks)*

   (Allow $100 \times \dfrac{14}{132} = 10.6$ for 1 mark)

5 a) i) mantle / asthenosphere (not magma alone) *(1 mark)*

      ii) e.g. inner core solid and outer core molten / inner core more dense / inner core hotter / inner core has more iron / inner core has more nickel (do not accept answers about size) *(1 mark)*

   b) i) (part b should read as a whole so marks can be awarded in either section)
   from molten rock / magma / lava *(1 mark)*
   forced / rises to surface / cools / solidifies / crystallises *(1 mark)*

      ii) slow cooling creates large crystals and vice versa *(1 mark)*

   c) **EITHER:**
   plates move apart *(1 mark)*
   magma / lava rises / fills gap or volcanoes *(1 mark)*
   OR
   plates move towards each other *(1 mark)*
   one plate driven down to form magma / volcanoes /subduction zone / magma formed *(1 mark)* (up to max of 2 marks)

6 a) alkanes *(1 mark)*
   It's not an alkene because there are no double bonds. It can't be a carbohydrate or a carbonate either, because there's no oxygen. So it must be an alkane.

   b) In plentiful / excess air *(1 mark)* carbon dioxide produced *(1 mark)*; in limited air *(1 mark)* carbon monoxide / carbon / soot produced *(1 mark)* (2 marks for products, 2 marks for distinction in the amount of air. Any reference to incorrect product e.g. hydrogen, ethene loses 1 mark)

   c) i) Cracking / thermal decomposition (accept decomposition, reject endothermic / exothermic) *(1 mark)*

      ii) $4C_2H_4$ *(1 mark)*
      H_2 *(1 mark)*

      iii) **from** red / brown / red-brown / orange / yellow *(1 mark)* **to** colourless *(1 mark)*
      (**Reject** 'clear' / discoloured. **Accept** 'decolourised' / paler)

   d) i)

   (1 mark)   Indication of long molecule *(1 mark)*.
   Single bond between carbon atoms *(1 mark)*
   The double bonds in the ethene molecules open up, so that they can link together to form long polyethene chains. You don't have to draw huge, long molecules though — just draw one section and show it's repeated lots of times by adding brackets and 'n'.

      ii) Does not tear / waterproof / airtight / no trees cut down / easy to make into a film / transparent / does not rot *(1 mark)*
      Does not rot away / uses up oil supplies *(1 mark)* (does not rot cannot be credited twice)

7 a) 40°C *(1 mark)*
   Enzymes like it warm, but not too hot — in fact, temperatures above 45°C damage most enzymes.

   b) New substances / products formed / lactic acid formed / mixture goes solid / starts to thicken / change in smell or taste / energy change / exothermic (not endothermic) *(1 mark for any one)*

   c) Protein / biological catalysts *(1 mark)*
   speed up reaction / alter rate / decrease activation energy *(1 mark)*
   (total of 2 marks)

8 a) i) Alkali metal **X** — rubidium / potassium *(1 mark)*
   Gas **Y** — hydrogen *(1 mark)*
   Solution **Z** — potassium hydroxide / X hydroxide *(1 mark)*
   (X **must** belong to group I or II)

      ii) turns purple / blue *(1 mark)*
      Z is alkali / alkaline / pH>7 / basic *(1 mark)*

b) i) 11.2 (g) *(1 mark)*

ii) $\dfrac{100 \times 11.2}{20}$ (=56%) *(1 mark)*

You don't need to know any chemistry for this bit — it's just maths.

iii) acid — sulphuric (acid) *(1 mark)*
alkali — sodium (hydr)oxide *(1 mark)*

c) i) boron / carbon / nitrogen / oxygen / fluorine / neon *(1 mark)*

ii) beryllium *(1 mark)*

d) i) electron arrangement 2, 1 *(1 mark)*

The atomic number gives the number of electrons — so just fill up the shells from the inside.

ii) Outer shell *(1 mark)* has same number of electrons / one electron *(1 mark)*

9 a) i) Energy required to break the bonds of both reactants:
436 + 243 = 679(kJ) *(2 marks)* (one mark for correct method)

ii) Energy released when the bonds of the product are formed:
2 × 432 = 864(kJ) *(1 mark)*

b) The overall reaction is exothermic. *(1 mark)*
More energy is released by bond formation than is needed for bond breaking. *(1 mark)*

c) A catalyst reduces the minimum initial energy required by reacting particles for the reaction to occur (the activation energy). *(1 mark)*

10 a) i) A getting larger and B getting smaller *(1 mark)* (Accept heavier/lighter, thicker/thinner, B disintegrates, reference to anode/cathode or copper from B to A).

ii) Electrolyte / provides (moving) ions / solution for the ions to go through *(1 mark)*

iii) $2e^-$ *(1 mark)*
Cu^{2+} *(1 mark)*

b) Good conductor of electricity, does not corrode, bends, ductile *(any two for 2 marks)*
Doesn't react with water, bends, good conductor of heat *(any two for 2 marks)*

SECTION B

11 a) i) Weak acids are only slightly ionised in water. *(1 mark)*

ii) Test the pH of aqueous solutions of citric acid and hydrochloric acid using a pH meter or Universal Indicator paper. The pH of the weak citric acid will be 4-6 whereas the pH of hydrochloric acid should be 0 or 1. *(2 marks)* (1 mark for knowing to carry out a pH test, another for knowing the pH of the different acids)
Or: See how fast a sample reacts with e.g. magnesium. *(1 mark)* The hydrochloric acid will react more quickly than the citric acid. *(1 mark)*

b) Arrhenius and Bronsted-Lowry thought that acids released H^+ ions in the solution. *(2 marks)* Citric acid behaves as an acid because it releases H^+ ions in a solution. *(1 mark)*

12 a) i) magnesium, copper *(1 mark)*

ii) furthest apart in reactivity series *(1 mark)*

b) concentration / temperature *(1 mark)*

c) i) $Zn \rightarrow Zn^{2+} + 2e^-$ *(1 mark)*

ii) zinc loses electrons more easily than copper / zinc is more reactive than copper so electrons flow from zinc to copper *(2 marks)*

13 a) cream (or similar colour, **not** white or yellow) *(1 mark)* precipitate *(1 mark)*, silver bromide *(1 mark)* *(any two for 2 marks)* (ignore 'spectator' ions)
$Ag^+ + Br^- = AgBr$ *(2 marks)* (-1 per error)

b) i) **J** is iron(II) hydroxide *(1 mark)*

ii) **L** is iron(III) hydroxide *(1 mark)*

iii) chlorine is an oxidising agent *(1 mark)*; any **one** point about change involving chlorine or Fe^{2+} to Fe^{3+} *(1 mark)*

14 a) An explanation to include:
bonds broken: none *(1 mark)*
bonds formed: H—O bonds / water **only** *(1 mark)*

b) A description to include **four** from:
burette **and** pipette
acid in burette / alkali in pipette
suitable named indicator (**reject** universal indicator)
correct alkali $\rightarrow$ acid colour change (methyl orange: yellow $\rightarrow$ orange; phenolphthalein: red $\rightarrow$ colourless)
further relevant point *(4 marks)*

c) A calculation to include:

amount of acid = $\dfrac{20}{1000} \times 0.5 = 0.01$ (mol) *(1 mark)*

amount of alkali = amount of acid *(1 mark)*

concentration of alkali = $0.01 \times \dfrac{1000}{25} = 0.4$ (mol dm^{-3}) *(1 mark)*

(accept 0.625 with some working shown for 2 marks)
For this one, you're going to need the formula:
number of moles = concentration × volume
Beware of units too — it's no good if the concentration's given in mol dm^{-3}, but you leave the volumes in cm^3.

d) A calculation to include:
NaOH = 40 *(1 mark)*
concentration of NaOH = 40 × 0.4 = 16 (g dm^{-3}) *(1 mark)*

mass of NaOH = $16 \times \dfrac{250}{1000}$ = 4 (g) *(1 mark)*

(Allow error carried forward from part (c))

Index

Index

Index